Surfaces:
Explorations with Sliceforms

Written and Illustrated by

John Sharp

**QED Books
an imprint of**

Tarquin Group
www.tarquingroup.com

Publisher's Note

All Trade Marks in this work are hereby acknowledged.

© John Sharp, 2004

ISBN 9781858532011

Tarquin Publications,
Suite 74, 17 Holywell Hill,
St Albans AL1 1DT,
United Kingdom
www.tarquingroup.com

A catalogue record for this book is available from
the British Library.

Cover design by Jane Conway

Printed in the UK and printed and distributed by IPG Books in the USA.

Contents

Contents

Preface

Geometry is a part of mathematics that does not belong to mathematicians. It is part of all cultures in defining aesthetics from patterns through architecture to other aspects of day to day design. The popular view of mathematics is that it is not concerned with spatial/ visual ideas, so that most non-mathematicians talk about geometry and mathematics as two separate subjects. I have had many students say that they were no good at mathematics, but liked and got on well with geometry. They see mathematics as an impenetrable collection of logic and formulae. This is partly the result of the decline of geometry during the twentieth century as an important field of mathematical research, but it has not died within our general culture because it is essential to so many human activities.

Geometry had a revival during the nineteenth century. Fermat and Descartes had devised a means of representing geometry algebraically in the seventeenth century and swung it away from its visual aspects. Despite the geometrical revival in the nineteenth century, most books on geometry could get away with few illustrations. However, in the latter half of the nineteenth century as the mathematics of surfaces was gradually unravelled, surfaces came to be modelled physically as well as algebraically. If the computer power we have now had been available, then who knows what direction mathematics would have gone.

The Victorian era was one of great advances in technology and the Great Exhibition of 1851 in Hyde Park in London was the precursor of many other such Expositions where the fruits of art and technology were displayed together. The Great Exhibition made a profit which resulted in a charity which is still in existence. The area of London known as South Kensington is its legacy. This included the South Kensington Museum which split into the famous Victoria and Albert Museum (known colloquially as the V&A), which specialises in the decorative arts and into the Science Museum. In 1876 the South Kensington Museum put on a huge exhibition of scientific equipment by collecting together instruments and models from all over the world. This included about 200 mathematical models and instruments. In the handbook to this exhibition, H J S Smith, Savillian Professor of Geometry at Oxford, said:

"So great has been the influence of the Cartesian mode of representation upon geometrical speculation that it has perhaps, to a certain extent and in certain cases, unduly led away the minds of geometricians from that direct intuition of space upon which geometry must after all be founded. And there can be no doubt that an Exhibition of models such as those included in the present Catalogue is calculated to render a great service to geometrical science by calling attention to the concrete shapes of objects, which are too apt, even in the mind of serious students, to exist only as conceptions very imperfectly realised."

This sentiment was also echoed by one of the giants of nineteenth century geometry Felix Klein, whose "Erlangen Program" gave a new form to geometry as the study of the invariants of a group of transformations. Klein believed that it was essential to have an actual mental image of the entity being studied and both made plaster models and encouraged his students to do so. One of these, Walther von Dyck produced a catalogue of models and devices for the newly formed German Mathematical Society in 1892 which ran to over 400 pages. Many of these models were exhibited at the World's Fair in Chicago in 1893 and were purchased by universities and colleges throughout the United States. They still languish hidden away on their shelves or in cupboards and in the storerooms of museums like the Smithsonian. They also exist in Europe and a set of photographs with a mathematically commentary was collected by Gerd Fischer in 1986 (see Fischer).

There were a number of companies producing models which seem to have been taken over by the Leipzig firm of Martin Schilling whose catalogue expanded until 1911 when it contained over 40 series and almost 400 models and devices. But the great period of model building ended during World War I. This was partly due to economic reasons with models cost typically £200 ($300) to £400 ($600) at today's prices, and that the pendulum of geometry was starting to swing away from the visual back to the algebraic. Klein died in 1925; his influence waned before this and he had already pointed the way to a study of the general rather than the specific.

There was one aspect that did not die, the visual impact was passed on to artists and continues to the present day. The artist Naum Gabo and other constructivists saw the models in museums. Man Ray photographed a number of them in the Instituit Poincaré as *objet d'art* when they were discovered there in the 1930s by Max Ernst. (see Wieder).

One set of models, of quadric surfaces, made by slicing the surfaces, is shown at the beginning of the book. They are the ancestors of Sliceforms. They now form part of an exhibit in the Science Museum in London called "Strange Surfaces" alongside a selection of my Sliceforms. In Martin Schilling's catalogue, they were shown as the engraving reproduced in figure p-1, which is hard to understand unless you can identify the models.

figure p-1

The models were invented by the mathematician Olaus Henrici , a Dane who came to England after studying at various places in Europe and were manufactured by the firm set up by the mathematician Alexander Brill in Darmstadt and described in the Catalogue of the 1876 exhibition as follows:

"These models are distinguished from those in common use by their mobility, by means of which each one represents not only a single ellipsoid or hyperboloid, but a series of surfaces of one or the other kind. For when the angle of inclination of the circular sections is altered, in a direction easily recognised by pressing or drawing out the model, there will be obtained a simple but infinite system, individual forms of which can be converted from a flat figure through gradually-changing solid bodies to just such another figure with a different relation of axes, without, however, losing its properties."

Whereas artists like Naum Gabo used the method of making the model for creating works which were essentially static, I was attracted by the dynamic aspects described in the above quote in such a typically verbose Victorian way, and their artistic impact is the origin of this book. Although I began my career as a chemist, I was always interested in mathematics as a recreation, encouraged, as were many of my generation by Martin Gardner's column in the Scientific American. I was also interested in art, and being colour blind, was particularly interested in shape and form and illusion. I began teaching, in

continuing education, in London in 1978 in classes which evolved into the
area of geometry and art. My modelling was originally polyhedra. I discovered
the technique which was eventually to become Sliceforms in Cundy and
Rollett's book of mathematical models and developed it in order to teach three
dimensional spatial awareness. It then almost became an obsession as more
and more ideas came my way to develop it and I delved into the history of
making models. Whereas recreational mathematics is often seen as concerned
with creating and solving puzzles, I have found that the challenges of making
the models described here have been more interesting.

There are many models in the book and they have been developed over at least
fifteen years, with periodic episodes of making a set of models as the creative
ideas came to me, or where I needed to produce an example for teaching. They
have taken over at various times since once I could see the model in my mind,
I wanted to see it for real and one model gave me ideas for more. When I have
wanted to see what a particular model was like, it has been an obsession at
times to have it realised.

It is said that the French mathematician Henri Poincaré (1854-1912) was the
last "universal" mathematician. He wrote that "A scientist worthy of the name,
above all a mathematician, experiences in his work the same impression as an
artist; his pleasure is as great and of the same nature". This sums up another
aspect of making models of Sliceforms, the balance of the beauty of the
mathematics with the aesthetics of the result. E T Bell (Bell) says that Poincaré
was an intuitionist and having once arrived at a summit, he never retraced his
steps. "He was satisfied to have crashed through the difficulties and left others
the pains of mapping the royal roads destined to lead more easily to the end."
This book is in that spirit in some ways; because I have concentrated on the
techniques and geometry of *making* a wide range of models, I have not
described the mathematics of their significance too deeply. I have described
new surfaces (especially in chapter 9) or methods for creating ones which have
not been modelled as far as I can see; there is need for more able
mathematicians than myself to take them further.

The reference in the Poincaré quote above to the "royal road" also has other
connotations in making the models. The origin of the phrase is with the
mathematician Menaechmus (c 375BC to 325BC) who was tutor to Alexander
the Great. When Alexander requested him to make his proofs shorter, he is
reply was that "even through there were private roads and even royal roads, in
geometry there is only one road for all". Some of the models can be designed

quickly, but many might take a day to design. Alexander the Great did take the shortest path where possible, like in cutting the Gordian knot, and the equivalent of that is to use a template and make the model from that.

The reader is thus warned that designing some of the models is going to take time, and that patience is a requirement in most cases. Assembly can take from a quarter of an hour where there are only a few slices with an hour more commonly the case (and ten hours with one or two of them). As with any skill, the more it is practised, the easier it becomes. I have described some models with reduced numbers of slices. This can give a model which has different properties and the simplicity is a virtue in its own right. I have taught people with a wide range of ability and presented workshops at Mathematics Fairs where all age ranges have come and cut out from pre-prepared templates. It has amazed me how many young children have been able to work out how the slices fit together to form the models (as well as how skilled some young children can be when cutting out while other older ones could not cut a straight line). Some people at these fairs have spent a large part of the day making the models. I do not think it is just the beauty of them, or that they had something to take home at the end of the day, but the sheer sense of achievement in solving the puzzle of the assembly. At one fair in York in October 2000, at least 1000 models were made in a day. I find it relaxing to sit and construct a model and some people find them therapeutic in helping them to unwind. I find it soothing to listen to music or the radio while making them. I also know someone who made a set of them while she sat with her father as he died over a number a days. Some teachers have also told me that where they have used sessions in making the models they have never known their pupils so quiet and concentrated so much on what they were doing.

Although this type of model has a long history, its full potential has not been realised. The methods described in this book are a starting point for mathematical and artistic exploration, so background on how the models were constructed and design hints are included. You do not need a mathematical background to understand the models themselves. If you are designing your own models, and are not a mathematician, many of the methods just rely on a sense of three dimensional space. They are also an excellent means of developing such a sense. To show that such models can have a practical use, in a rigid form rather than the flexibility of Sliceforms, look closely at figures p-2 and p-3. They are a drawing of a plan for a ship taken from Thomas Bradley's "Elements of Geometrical Drawing" published in London in 1862, showing

the stern and part of the plan of a paddle steamer called the Nile. They show
how the structure of the surface of the vessel is defined as a set of slices.

Stern Elevation

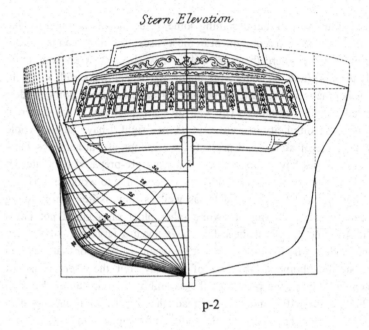

figure
p-2

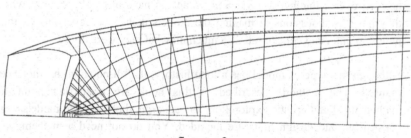

figure p-3

There is a philosophical question that surfaces regularly which asks "Is
mathematics discovered or do we invent it?". It is unanswerable, of course, but
my aim in writing the book is to make it more of a puzzle by showing that, as
far as the geometry of surfaces goes, it is both possible to be a discoverer and
an inventor. To do so is to gain a sense of the aesthetics of mathematics as
well as having great fun.

Ways you might use this book

Sliceforms mean different things to different people. Your interests may be quite specific. For example if you want artistic/sculptural ideas, you will probably find the chapter on surfaces from equations too mathematical. However working forwards in the book will provide an invaluable insight into the nature of surfaces.

The following description is a guide to the book, from a number of different approaches, with a summary of what is in the chapters and help in identifying where to find ideas for your needs.

Chapter summary and the book's structure

The chapters to some extent progress in complexity, but that does not mean some cannot be skipped since the way the initial techniques are developed depends on the type of surface being constructed. Most chapters have a structure that takes you through the basics and then develops its ideas to create new Sliceforms. This includes examples, but mostly as a starting point for your own original, creativity.

Preface

This gives a history of the subject and background to both its geometrical and artistic nature. You will find this useful if you want to know how the book came about and why the subject is important for both artists and mathematicians.

Introduction and basic techniques

This is the chapter that you must read if you are to understand the tricks of making Sliceforms as well as the fundamental ideas behind them.

Quadric surfaces

This chapter starts from the sphere and shows how to build more complex surfaces, using the simplest curved surfaces derived from the conic sections. You will find this chapter useful if you want to move on from the ideas of the introductory chapter, but also to understand some of the subtle concepts of the geometry of slicing a surface. Ideas from this chapter are developed in the chapter on *Variations and explorations (see below)*.

Contours and profiles

Geometry is a tool for many subjects outside science and mathematics. This chapter approaches surfaces from the point of view of contour maps and surfaces defined by profiles. You will find this chapter particularly useful if you want to make geographical models or are sculpting surfaces which you intend to copy as profiles.

Properties and types of Surfaces

You do not need to understand the basics of how surfaces curve and how they can be classified to undertake many activities with Sliceforms. However, reading this chapter may help in deciding how to make different types of surfaces. It also acts an introduction for the next three chapters. No construction methods are presented and, as such, it could have been part of the introduction. But you will appreciate the finer points of the properties of surfaces if you read it having worked through the initial chapters.

Surfaces of revolution

A common method of generating surfaces is to take a curve and rotated it about an axis, hence "surfaces of revolution". This chapter shows how surfaces of revolution can be created as Sliceforms. More ideas are developed in the chapter on *Variations and explorations* and surfaces of revolution are also treated algebraically in *Surfaces from equations*. You will find this chapter useful for general ideas about the geometry of such surfaces, and also for artistic inspiration.

Ruled surfaces

Many surfaces can be defined as the path of a moving straight line. As with the previous chapter, after exploring the basics of such surfaces the chapter goes on to explore new ways to make them. There are also ideas for adapting the way the slices are drawn so that some Sliceform models can be made in ways which change their appearance and the way they behave.

Surfaces from equations - Algebraic surfaces

As the title suggests, the contents of this chapter are more mathematical. This means the chapter will be very useful for mathematics teachers or anyone interested in understanding different types of equations and how they can be solved to create a range of novel and beautiful Sliceforms. Some surfaces require special techniques which are described in the *Computer* chapter (*see below*). As well as describing methods for handling different equations, it also shows how to be creative with equations so that new surfaces can be invented. So if you are an artist looking for new ideas, and you feel the ideas are out of your depth, look at the pictures and consider finding someone who can help with the mathematical and computer side; such collaboration is in the spirit of how I came to write the book.

Polyhedra

Polyhedra are often considered as solids rather than surfaces since they are made up of flat polygons, and they are familiar as solid models in everyday life and in nature. Sliceform models of them allow them to be seen in a different way, and open up the possibility of discovering new properties. They can be tools in mathematics teaching, but equally well they suggest new ideas for the artist.

Variations and explorations

In this chapter, many ideas for making the Sliceform models discussed in previous chapters are developed. It is about being creative with different methods for making models. These may be artistic explorations, but there are also ideas which may offer new mathematical insights. While there are descriptions of different models, the chapter is primarily intended to spark creative ideas for anyone.

Using the Computer

This chapter covers the use of a computer in designing Sliceforms for a wide range of users, so you need to look though it to decide what level is suitable for your knowledge of computer techniques. It covers the various types of drawing software all the way up to general programming and ideas for particular types of Sliceforms.

How will you use Sliceforms?

Sliceforms have applications in many subject areas. The following are a few of these and the most important chapters for each one. However you want to use the book, make sure you are thoroughly familiar with the ideas in *Introduction and basic techniques*.

Artists and designers in three dimensions

Sliceforms can be seen as an example of where art and geometry meet, but they also have other uses for artists. Normally, they are dynamic objects but they can also form a rigid sculpture. They influenced Naum Gabo to create his series of heads, but they can also be used to model and then completed using a medium like plaster to give a conventional solid surface. They can also act as possible inspiration for three dimensional textiles. An example is shown of a chair design in the chapter *Variations and explorations* (*see above*). They are also ideal for learning how to handle three dimensions, since they involve geometrical reasoning as well as construction.

If this is how you want to use the book, then once you are familiar with the *Introduction and basic techniques* choose another chapter and find the method that you wish to use. *Variations and explorations* has many ideas as starting points for new ways to work.

Mathematics Education: All levels
The geometry of Sliceforms offers many routes for providing a resource for teaching. The design and construction methods are rich ways of learning about three dimensions, encouraging thinking and producing something very aesthetically pleasing as an end result. To have an object to talk about enhances attention and Sliceforms have the additional benefit that they are portable and easy to store.

Mathematics Education: the Primary and Middle School
At level of primary education, they offer a fun way to explore three dimensions and ask questions about surfaces and polyhedra. They can also act as additional source of examples, such as exploring symmetry. The early chapters and the *Polyhedra* chapter may be the most useful ones here.

Higher levels
At higher levels the whole book offers many ways to provide meaningful exercises with a very attractive end result. The chapter *Surfaces from equations - Algebraic surfaces* has some ideas for developing thinking on using equations in a novel way, and there are many other ideas in the chapter on using the computer.

Engineers and Architects
At one level, the book can be seen as a source of inspiration, but it can also be used to teach three dimensional thinking. Some of the surfaces such as the hyperbolic paraboloid in the chapter on *Ruled Surfaces* are familiar as building structures. The static shapes of the slicing of surfaces suggest the floors of buildings.

The simplest rectangular shapes have been used for packaging for a long time, and they have the added benefit of being easily storable, before and after use. The way Sliceforms behave dynamically has been used as a form of linkage mechanism, but there is still potential for more three dimensional exploitation.

Making the models

While the illustrations in the book are necessary to understand the models and their design, at one level I hope you will look and wonder at the beauty of the mathematics they reveal. But I also hope that it is more than a coffee-table book of mathematics. The only way to appreciate the fully beauty of the models is to make them, hold them, turn them around and open and close them. They are dynamic and not static.

The main construction techniques are described in *Introduction and basic techniques*, but other special methods are scattered throughout the other chapters.

Before designing models of your own, I would suggest that you make some models, first by copying the templates at the end of the book (or from the internet as described in the section at the end of the book on *Where to find out more*) onto card and then cutting them out and assembling them. Apart from the satisfaction of producing a beautiful object, you will find they can help you develop your three dimensional skills through their puzzle aspect.

Creating your own models, or even just following the techniques helps in understanding the mathematics behind various surfaces and solids, as well as developing three dimensional thinking and spatial awareness. If you are teaching mathematics, even to have the models for discussion of the object in question proves helpful. They also raise questions in a range of ways.

If you are not a mathematician there is scope for learning techniques which are useful for art and design. If you see any mathematics you do not understand, then skip over it and use the descriptive geometry.

When constructing your own models, I would advise drawing the slices using a computer. This has the advantage that you can copy slices where you need duplicates and you can print multiple copies.

A word of warning

The sense of satisfaction in making a Sliceform model is addictive.

Chapter 1
Introduction and basic
techniques

This chapter describes the basics of creating and assembling Sliceform models and should be read before attempting any models in other chapters.

What are Sliceform models?

The pictures of Sliceform models in this book can only give a hint at their beauty because they are *dynamic* models of surfaces or solids. They are designed by *slicing* objects at regular distances in two directions, so that looked at from above the slices fit on a grid. A slot is cut up or down from the points of intersection of the grid on each slice. The slots are normally cut so that they all start from the base of slices in one direction and from the top in the other direction. Figure 1-1 shows a pair of slices and their slots and the two slotted together.

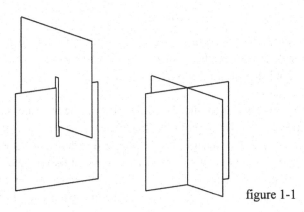

figure 1-1

When the object is reconstructed by fitting the two sets of slices together, the slots act like a multitude of hinges which allow the object to flex and *transform*. The model then collapses flat in two ways but in intermediate positions goes through a host of different but related shapes. This ability to collapse flat gives them an edge over many other models, since they are easy to store or send through the post.

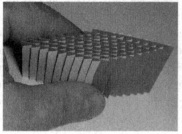 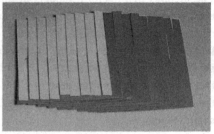

figure 1-2

The name Sliceforms has been coined to describe that the models are made from a set of slices and that they transform. This name was invented by Gerald Jenkins, the publisher of my first book on Sliceforms.

By using different colours for the slices, especially if the two sides of a slice in each direction are different colours, the patterns generated as the slices obstruct one another also change. The colours and patterns also change as the model is rotated. If you look in the direction of the grid forming the slices, you can look through them. Tilt them slightly and the interplay of the colours and shadows of the slices gives them a magical quality. Hold them in front of a light and they cast a multiplicity of shadows, which vary not only as the model is rotated as shown in the following illustration, but also with various stages of flexing (flattening).

figure 1-3

You do not need a mathematical background to understand the models themselves. If you are designing your own models, and are not a mathematician, many of the methods just rely on a sense of three dimensional space. They are also an excellent means of developing such a sense. The emphasis is on geometrical techniques and exploring the objects visually and with the hands, although there is some mathematics for anyone who is interested in that side of the subject.

Models can be made from many different kinds of solid objects and created in many different ways, as the rest of the book shows. This chapter, while describing the design and construction of actual models, mainly deals with models based on the cube. The next chapter moves up to surfaces which historically were the original Sliceforms created in the nineteenth century. The rest of the book develops ideas in different ways.

First steps in designing and assembling models

The easiest Sliceform model to design is a cube, for the simple reason that all slices can be made the same. The following descriptions of how to make one will help you to see basic principles both for design and construction.

Materials

I use card suitable for photocopying or for use on a laser or inkjet printer which is sold for creating covers for reports. Its is also often sold in small packs in stationery or craft shops, for making cards or general craft work. It needs to be 160-180 gsm (60-65 lb) weight. The advantage of using such a card is that a drawing can be produced and then multiple copies made without destroying the original, even if you are drawing by hand. I draw designs almost exclusively using a computer and need to print directly. You need to ensure that your photocopier or printer can handle this card, but almost all do.

The finished models are usually 5 to 15 centimetres in each direction. I have made larger models in thicker card and plastic that are more rigid and have seen one in wood which is about a metre high, but this presents other problems. With photocopy card, there is an art to cutting the slots (see figure 1-8) but card is tolerant of a wide error. Thicker card, plastic or wood is not as easy to work. The initial problem of transferring the design adds to the time taken to make the model. This is not always easy and with thicker material, accuracy, both of design and transfer, is more crucial. Models made with thicker material usually do not collapse and transform. With wood or thick plastic the strength of the material can also mean a health hazard. Some slices have very sharp points at the slots as shown in figure 1-4. Take great care when assembling such models as you can easily stab yourself if the slice slips during assembly.

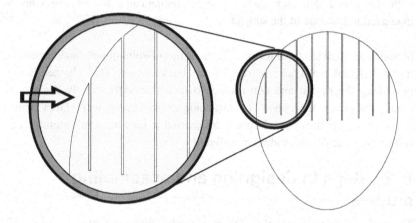

figure 1-4

The amount of card and number of slices required can sometimes be deceptive. To bring the solid volume and area required into perspective, think how much area a loaf of sliced bread will cover when the slices are spread out. Choosing nine slices in each direction and aiming for models around 5 to 15 centimetres in each direction means the slices for each direction can fit on a single sheet of card. This also matches the physical properties of the card, since larger models made on card of 160 gsm weight do not hold together so well.

In the following example, a pencil, paper and ruler are all you need to draw the slices. I use a computer almost exclusively since I can scale the model, or print out multiple copies. A separate chapter has been included for computer production of models describing techniques and the types of software to use.

Designing a cube with a square grid

The starting point for many constructions is the grid which defines the positions for the slices. In most of the models in the book a square grid has been used. As the model is flexed and collapsed, so the grid distorts and becomes a grid inclined at a range of angles; the shape of the Sliceform also changes. You can use other grids which are inclined at various angles for the initial design, but you will get a different model. This seems paradoxical, but because the slices will be of different shape when you design using a different grid, and so the transformations and the position which restores the original object will not be the same. Chapter 9 describes variations using different grids and how to create new models from ones you have already designed.

Stage 1 - the grid

The first stage in constructing the cube is to draw a plan of the grid as at the left of figure 1-5.

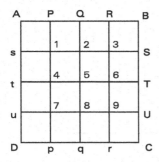
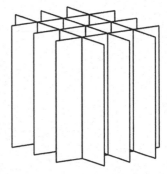

figure 1-5

The base of the cube has corners ABCD. The two sets of slices are referred to by the conventional cartesian directions of x and y. Thus the slices which fit above the grid lines sS, tT, uU are the x-slices and those above the pP, qQ, rR grid lines are the y-slices. The lines AB, BC, CD and DA cannot be used as slices, since they have no support. This means that the Sliceform cube at the right of figure 1-5 has the surface on the top and base more complete than the side faces. This mixture of different surface appearances is more obvious in Sliceform models of objects with flat surfaces. Where the surface being modelled is curved, as with many models in the book, the effect is only there if you look for it.

Note that there are three slices in each direction and consequently the central slices go through the centre of the top and bottom face of the cube. Having an odd number of slices is important in most models for this reason. See the section *The number and position of slices*, later in this chapter for more details.

Stage 2 - design the slices

This is usually the most involved stage. However, in this case it is easy to see that each slice of the Sliceform cube is a square. It is very unusual to create a model which only has one type of slice, although the slices in the two grid directions may have the same shape.

When marking the position of the slots, use the grid to find their positions. For example the slices corresponding to grid lines pP and sS are shown in figure 1-6. The numbers correspond to the positions of the intersection of the grid lines in figure 1-5.

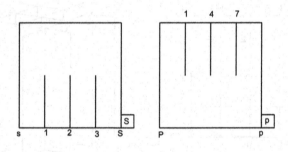

figure 1-6

As you can see here, there is another addition to the design: two labelled tabs at the bottom of the slices. It is a good idea to add these to identify the slices and their orientation. This helps when assembling the model. In this model, it does not matter whether you add them or not, since all slices are the same. In some models, there are slices which are only slightly different in size and when you have cut them out, it is easy to lose track of which slice is which. Cut them off when the model is assembled. The assembled cube of figure 1-5 looks like figure 1-7 when the tabs are still present.

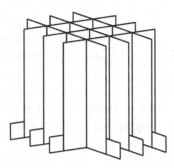

figure 1-7

Another way to identify the slices is to mark identification for the slice next to the slice on the card. This method is best if you are cutting and assembling one slice at a time. If you are cutting all slices out and then assembling (as might happen where cutting out is shared by a group of people) then the tab method is best.

The slices in figure 1-6 are a typical x-slice and y-slice and so have the slots in different positions. The length of the slots are measured as follows. The uncut part of the line above point 1 of the grid at the left of figure 1-5 of slice pP matches the slot length of slice sS and vice versa. In principle only the length of the cut part of each slot needs to match the uncut part of the slot on the other slice, but as a general rule, measure half way. You then always know that the slot will fit. Since there are many slots to match, it is easy to run into problems if this rule is not followed. Some models require a great deal of effort in design and the frustration of your trying to sort out problems caused by using different slot lengths is just not worth your trying to be different! It is also takes much longer to design a model if you have to make extra decisions about lengths.

Stage 3 - cut out the slices assemble the model

There are a number of ways to approach this stage:
- cut out all slices first, then assemble them
- cut out a pair of slices, one from each direction and add them to the model
- cut out slices in one direction then cut out the other direction slices and add them one by one
- cut out with scissors, or cut out with a craft knife
- assemble from the middle outwards, in pairs in each direction
- assemble all slices in one direction into the central slot of the other direction, then add the rest of the slices
- in a few cases it is easier to assemble from one end.

Whichever approach you choose, **the most important point to bear in mind is how you cut out the slots.** I have seen many models ruined when this point has been ignored.

My preference is to cut out pairs of slices with scissors and assemble from the centre outwards. The reasons for saying this are that I find it breaks up the type of task you are performing and you see the model growing earlier. Even if I add tabs to the model (and I usually do not, preferring to mark the number of the slice next to the slice pattern) it also helps me to keep track of the order of slices. Some people prefer to use a craft knife, but most models do not have straight edges to the slices and so a special craft knife with a rotating blade is required to cut these. I also prefer to sit in an armchair, either watching television, or listening to the radio or a favourite CD, and there is no alternative to scissors in this case.

I will not describe cutting out around the edge of the slice, since it depends on the tools you choose to use. Once you have a slice ready to cut out the slot, you need to be aware of the following rules:
- the slot requires **two cuts**
- the width of the slot should match the thickness of the material you are modelling with
- the slot should be straight and not V-shaped.

The first point is the one you ignore at your peril. The reasons for this are obvious if you are working with large models made from planks of wood, say. If you do not make a wide enough slot, then the slice in the other direction will not fit. However, with paper or card if you make one cut, to give a slit rather than a slot, two slices can be fitted together but there is no hinge effect along the slit. The forces acting sideways on the slices will cause the model to buckle if you join two slices with only a slit cut in them.

Cut the slot with scissors making a pair of cuts either side of the line marked for a slot as shown exaggerated by the dashed lines in figure 1-8. Practice cutting slots using some spare card with a short pencil line drawn to the edge as the slots would be on the model. This will help you to gauge the correct thickness.

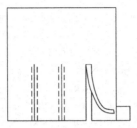

figure 1-8

The solid lines on the marks for the two left slots in figure 1-8 represent the line drawn as part of the design. The dashed lines represent the position for cutting the slot. *These positions are exaggerated to make it easy to see; in practice they will be very close to the original line.* When you cut along the two lines, this will give a sliver of card which will often curl up like a hair as shown in the cut slot at the right. Pinch it off at the end of the slot by grasping it with your fingernails. Remember to cut parallel to the marked line so that you get a definite width of slot at the top (as shown) and *not* a slot which goes to a point giving you a V-shaped slot.

When you have slices with all the slots cut, it is wise to check you have cut *all* slots because it can be irritating to have to disassemble the model to cut a missing slot.

Depending on the model, assembly can be easier when you start to assemble or when it is nearly complete. The order in which you assemble can affect this. The first method, to the left of figure 1-9 takes all the slices in one direction (the *x*-direction) and assembles their central slots into the slots of the central slice in the *y*-direction. In most models, this has the disadvantage that the slices in the *x*-direction are unstable and may drop out. Also, when you come to insert the remaining slices in the *y*-direction, you have to slot each one into the maximum number of slots in the *x*-direction.

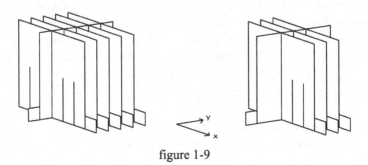

figure 1-9

I prefer the method whose first two stages are shown at the right of figure 1-9. First slot together the two central slices from each direction. Then slot the pair from the x-direction which are adjacent to the central slice of the x-direction. This is the stage reached in the figure. Now add the pair of slices in the y-direction which are next to the central y-slice. Continue adding pairs to move outwards in each direction until the model is complete. The advantage of this method is that the model assumes rigidity very quickly as the slices hold one another together by friction; the slots are less likely to fall out. Also, the task of slotting the slices together is easier because you are not necessarily fitting together the maximum number of slots each time.

Stage 4 - Consolidate the model

The final process is to make sure all the slices are firmly home and that the model folds correctly in both directions. If a model has a flat (plane) surface on which the grid sits, tapping the model gently on a table top helps to bed the slices and align them.

When you have all the slices fitted together, gently flex the model so that it folds flat. Do this slowly, because you may find that some of the parts of a slice either side of a pair of slots want to go the wrong way. Fold it flat a few times in each direction so that it is "run in". Do not worry if you hear a sort of crunchy sound; this is quite normal and happens as the slices flex over one another. Indeed, a model that does so is well constructed because the slots are cut just the right width to allow flexing.

Designing with non-square grids

Although many of the designs in the book are based on the slices fitting on a square grid, variations can be made using different grids. This should not be confused with the fact that the grid of a model becomes distorted as the Sliceform flexes. The volume of any Sliceform model varies as the model is flexed. However, two models of the same object can have the same overall volume and shape, but with the positions of their slices totally different. Apart from looking different, they behave differently as they are opened or flattened. One example follows, and others are described in chapter 9. Compare the Sliceform cube of figure 1-5 with the one shown in figure 1-12 to see the difference the slicing makes to the appearance.

Designing with non-square grids is not as straight-forward as using square grids. This example of making a cube with a non-square grid shows how the slices change shape and how care needs to be taken in matching the grid to the outline. The example also includes some pitfalls and shows how sizes sometimes have to be adjusted and how compromises have to be made.

The diagrams in figure 1-10 show two non-square grids placed on a square to find the size and position of the slices and slots. The version on the left is symmetrical, but suffers from the problem that the intersection of the grid falls on the perimeter lines (AB and CD). When you try to flex a model with slices made this way, it may not collapse easily. Models where the slices end in this way are not as aesthetic as ones where their slot positions are definitely away from the edge.

The design on the left has another problem which makes it unsuitable for a model. Slice xX (and its equivalents by rotation or reflection) have only one slot on them. This means the slice is not held firmly. Not only would it flap about as you moved the model, but there would not be enough frictional force to hold it in place and it would tend to fall out of the slot. If the square ABCD is made slightly larger, then the grid intersection at X would give another slot, but even then two slots are not much of an improvement. The diagram on the right shows how the grid has been shrunk.

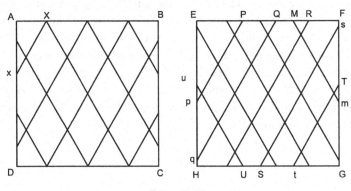

figure 1-10

This is now a compromise between missing the perimeter and getting a minimum of three slots. Further adjustment of the size of grids and the angle of the grid lines might improve this. The gaps at the corners are not ideal since they don't reflect the shape of the cube. When you look at a model the brain constructs its outline; it has less information to work with when looking at this model and you may not find it as aesthetically pleasing. If you are a perfectionist in this respect, it make take quite a while to strike the balance, since usually making changes to correct one type of problem results in another.

Once you are satisfied with your grid placing, you need to create the shapes of the slices. In this case, you know the height of all slices is the same as the side of the cube. The technique for positioning the slots is described in detail in chapter 3 (see figure 3-4).

The slices corresponding to the grid lines pP and qQ look like this:

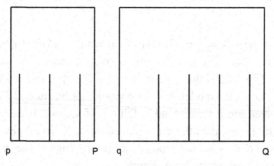

figure 1-11

and the slots have also been added by measuring half way up the slot lines.
In making the other slices, you can see that lines pP, tT, uU and mM are identical and similarly for other sets of lines. No slice has slots which are symmetrically arranged so there is a much more drawing required when making a model like this compared with the cube on a normal rectangular grid. Whereas the cube made with slices parallel to the sides gives the same result on folding in both directions, this model (figure 1-12) can take a square and rectangle form when folded.

figure 1-12

Other general design tips

Each chapter in the book describes particular types of surfaces or solids. Some of these lend themselves to being designed in a particular way. The following extra tips are points you need to bear in mind when designing any type of model.

Position of slots

When considering the position of slots you also need to take account of the fact that if a slot is close to the edge of a slice, then the strip from the edge of the slice to the slot may not be very strong. With models where the height of the strip is small (usually where the slice is highly curved) then this strip tends to get bent as you flex the model and can break off.

An example of this is slice pP in figure 1-11. The left hand slot is very close to the edge.

Directions for slicing

The cube examples above are easy to visualise because the cube is simple and you have a picture of the shape in your head and it is also highly symmetrical. When thinking about other designs, you also have to think about the slice in a vertical direction as well as the grid. In some cases there are many ways you can slice the solid and some of these may not be ideal. The direction of the slots may be crucial. The following slice shows slot lines placed in two positions. The top diagram shows slots placed conventionally.

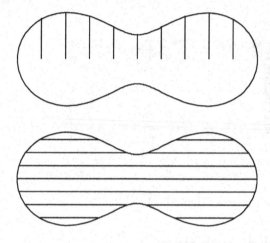

figure 1-13

The bottom diagram shows lines drawn in ready to halve to make slots. If you tried to slice as in this diagram, the top and bottom pairs of slices are in two parts. Avoid such cases wherever possible. Strategies for dealing with situations where a slice yields two separate parts are described in chapter 9. The slots are also very long in the central part.

Where you have designed slices in just one direction

There are many instances in designing models where it is easy to define the slices in one direction, but not in the other. This could be because the mathematics may be difficult or it may not be possible to get an accurate set of slices in one direction, for example with the profiles obtained using a sliding pin profiler as described in chapter 3.

If you have one set of slices and the slots are drawn on them, then you know the height of the slice at one particular place on the model in both directions. You can choose the spacing in the "other" direction, but the model looks best if the slots are the same spacing as on the slices you have already. This means that you can build up a set of heights for a particular slice in the "other" direction and freehand draw the shape of the slice. This method is described in chapter 3 in the section titled "When slices are only available in one direction" as part of the construction of a model of a cycloid surface.

The number and position of slices

When designing a grid, the position of the lines define where the surface is sliced. In many cases this can be crucial for the best model. It is best if a slice appears at all points of maximum or minimum curvature, or through any cusps. This can be a very subtle difference in some cases and often only apparent if you make two models and compare them. The following examples show important this can be, since features of the surface can be lost if they do not show on the slices.

For a symmetrical model, the central slices usually define a maximum or minimum curvature. Thus it is best that a slice appears at this position to define the surface fully and ensure it is smooth. Since the number of slices are generally symmetrical about the central ones, this means using an odd number of slices. This is illustrated by the following grids which are used to construct a sphere as described in chapter 2.

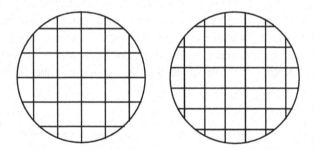

figure 1-14

The grid at the left of figure 1-14 is used to design a sphere with five slices in each direction and the one at the right with six. This means that the central slice of the five slice sphere will be a circle which has the same radius as the sphere. The one from six slices does not have a central slice and so the model has a flat area at the ends of the axis corresponding to the position of the point of maximum curvature. It is very difficult to see this result in a drawing or photograph (figure 1-15), since our brains are good at constructing the complete surface from a limited amount of information. However, if you make the models and try to balance the two spheres, then the six slice one will sit happily on this flat area and will not roll as well as the five slice one. You may want this ability to sit flat and not roll for a design which is functional in some way, but if you are designing for the best geometrical model, then you should aim for the five slice one (or one with a larger number of slices which is an odd number).

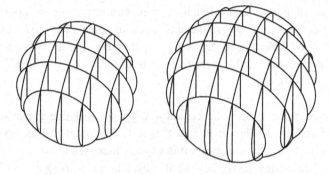

figure 1-15

It is much more important to consider how the slices are placed when there is a definite maximum or minimum, especially if this is a definite peak or valley in the surface. You need to look very carefully to spot the difference between the two models in figure 1-15 and would see no difference in individual slices because they are all circles. Figure 1-16 shows the slices from an equation plot of a version of the Cassinian surface described in chapter 7. These slices are markedly different. The correspond to a position where there is a pair of valleys.

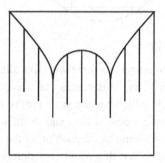 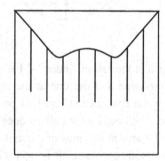

figure 1- 16

The slice on the left has been created with a grid having ten slices and the one on the right with one having nine slices. They are the slices which show the minima. In the model on the left the slice corresponds to the position of the cusps, whereas on the right the slice is just past the cusps and on the side of the valleys. In the slice on the left, note how the position of the third slot from each end is exactly on the cusp. Also, the central slot matches the maximum of the curve in the centre. In the slice on the right because there are an even number of lines on the design grid, there are a pair of slots either side if the maximum part of the central curve. This would lead to the same type of flattening as described above for the sphere. More importantly, the cusp which is very evident in the slice on the left would not be visible on the model. Such a Sliceform model where the cusps are in a hole of the model rather than on a slice would not be the best representation.

Avoid long slots

Surfaces which are roughly spherical or cubic make the best models, because this gives a good balance between the length of a slot and the width between them. Models which are very tall result in long slots and the mechanical properties of the Sliceform are such that the model becomes like a brush rather than a rigid object. When you try to flatten the model, the long strands tangle and get crushed. Such a model would be a cylinder where the slots are parallel to the axis of the cylinder. As shown in the next chapter, there are better ways to model a cylinder.

Chapter 2
Quadric surfaces

This chapter describes the construction of Sliceforms of the simplest surfaces apart from a plane, which are the surfaces equivalent to the conic sections in the plane. See chapter 4 for an overview of the geometry of surfaces.

The sphere

The simplest surface to model as a Sliceform is a sphere. Slicing a sphere in any direction always gives a circle, with the position of the slice defining the size of the circle. Circles are easy to manage in the design stage since the slots are drawn from a diameter to the circumference. Nearly all the slices for the surfaces modelled in this chapter are circles. The example shown in figure 2-1 could be from any one of the models.

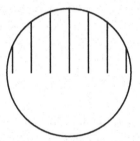

figure 2-1

There are a few properties that should be noted with this slice. Firstly, the slots are symmetrical; designing models so that this happens reduces the number of slices that have to be drawn. Secondly, the outermost slices are very close to the edge; this is a problem that frequently arises because the slope of tangent to the circle at the end of a diameter is parallel to the direction of the slots. It occurs in other models of surfaces where the slices are not circles, but, depending on the model, it may be possible to ignore such a slot. However, if you decide to do this, remember that you need to remove the slot from the slice in the other direction. Because of the symmetry of the sphere, this slice could have been from either of the sets of slices.

As described in chapter 1, different grids can be used to create different models. Figure 2-2 shows the grids for two different models of a sphere. In both cases, the radius of the circle has been adjusted to minimise ends of grid lines intersecting on the circle. The lines shown in each grid correspond to the diameters of the set of circles which make up the slices. The method for designing the slices is essentially the same technique described in chapter 1 for the cube, but using circles instead of squares. A more detailed general method is given later in the chapter, under the heading "The general method for construction using circles".

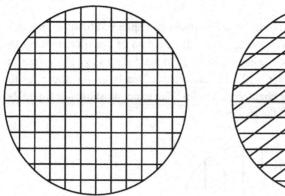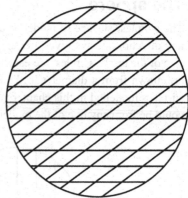

figure 2-2

The grid at the left in figure 2-2 has the grid lines at right angles. The grid goes through the centre of the circle in order to give a better sphere which has better curvature, for reasons described in the section "Odd or even number of slices?", at the end of chapter 1. There are only six different circles. Using a design with fewer slices, the different types of circle are shown identified as different types of line as follows:

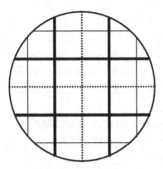

figure 2-3

Thus when designing the model, you only need to draw a limited number of circles and place the slots according to the position of the grid. Using the grid in figure 2-3, there are only three slices to design, and you need to create two circles for the central line and four of the others. Similarly, for the grid to the left of figure 2-2, there are six different circles to design.

The grid to the right in figure 2-2, however, does not have the same symmetry. Again, looking at a simpler version, with fewer slices in figure 2-4, there are a number of slices that are identical as shown with different types of line:

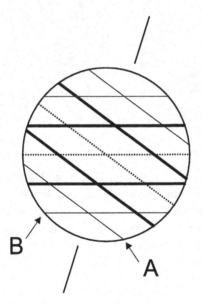

figure 2-4

The lines and hence the slices are matched by reflection about the axis of symmetry shown, so that slices at positions A and B are identical in shape (but mirror images on one another). There are only three types of slice to design, but since the grid does not consist of perpendicular lines, the slots are placed asymmetrically along the line. If you just make copies of the slices for each direction, when you assemble the model, you have to turn over some slices (such as the ones marked A and B) to make them slot into the correct position.

How do the two Sliceform spheres differ?

The two spheres produced from the grids in figure 2-2 have many different properties. For the purposes of the following discussion, the left one of figure 2-2 is the 90° grid sphere and the one at the right is the 36° grid sphere, these angles indicating the inclination of the lines of the grid.

The first obvious difference is that it is easier for the 90° grid sphere to be left standing as a sphere. The 36° grid one either feels it should flatten, or is more happy in the form shown in figure 2-8 which creates a surface known as an ellipsoid which is a position between being spherical and flat. Figure 2-5 shows photographs of them side by side as spheres.

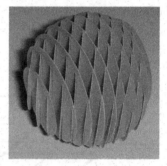

figure 2-5

The next difference is the way they deform. The 90° grid sphere, because of its symmetry, gives the same set of ellipsoids whichever way it is flattened. (See the section "Quadric Surfaces" below for a description of ellipsoids.) Consequently, the flattened form is the same in both directions, an ellipse. Figure 2-6 shows this ellipse flattened form.

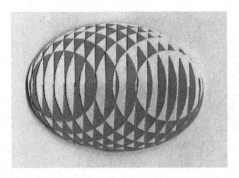

figure 2-6

The 36° grid sphere, however, is markedly different in its two flattened forms. In one direction, it is almost circular. In the other, it is a very long thin ellipse. In fact for spheres of the same size, the ellipse from the 90° grid sphere has a major axis which is less than half the one from the 36° grid sphere.

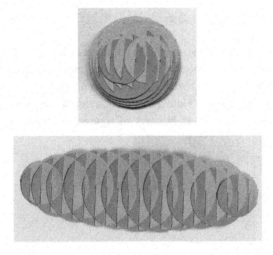

figure 2-7

The 36° grid sphere naturally wants to rest as an ellipsoid:

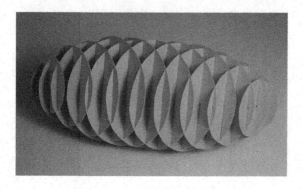

figure 2-8

Quadric surfaces

The sphere is one of a set of surfaces, the quadric surfaces. These are the three dimensional equivalent of the conics (the parabola, ellipse and hyperbola shown in figure 2-9) and their equations are all of the second degree, that is:

$$ax^2 + by^2 + cz^2 + 2fyz + 2gzx + 2hxy + 2ux + 2vy + 2wz + d = 0$$

where the coefficients define the shape of surface. Slicing a quadric surface results in a conic and quadric surfaces are sometimes called conicoids for this reason. The term conic is short for conic section, that is a curve derived by slicing or sectioning a cone. Note that a section through both sides of the cone produces the hyperbola (the right curve of figure 2-9) which is split into two parts.

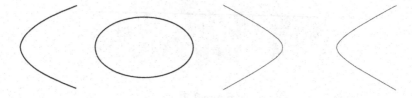

figure 2-9

In the plane, an equation of the second degree can give rise to a pair of straight lines as well as a circle, an ellipse, a parabola and a hyperbola. In space there are more possibilities. Some of these surfaces are shown in chapter 5 (figures 5-3, 5-4 and 5-5). The cone is a special quadric surface and just as slicing a cone in different ways produces different conics, so different conics *may* be produced by slicing other quadrics. The stress on *may* is because some quadric surfaces when sliced only give rise to one type of conic; for example an ellipsoid (see figure 5-4) can only give ellipses which includes circles as symmetrical ellipses.

Surfaces of revolution (see chapter 5 which shows diagrams of the conicoids) can be generated from each of the plane conics and to be a quadric surface they need to be rotated about the axis of symmetry of the conic. The surface of revolution of a circle about an axis through the centre gives a sphere, of course. An ellipse rotated about its major or minor axis gives a spheroid. A spheroid can be sliced perpendicularly to the axis of rotation to give circles. Other ellipsoids when sliced in a similar way give ellipses as slices instead of a circle. However, as we shall see in a moment, it is possible to slice all ellipsoids and get circles. A hyperboloid (with circular or elliptic cross sections) can be one of two types (of one or two "sheets" or "nappes") depending on which axis of symmetry is used to generate it (see figure 5-5) . A paraboloid can be circular (figure 5-3) or elliptic, and in addition there is a hyperboloid paraboloid (figures 6-7 and 6-8), whose slices are hyperbolae or parabolae but which is not a surface of revolution. The hyperboloid of one sheet and the hyperbolic paraboloid can be generated as a series of straight lines, that is a ruled surface; this method is described in chapter 6.

A cone is a degenerate case of a quadric surface where a point (the vertex of the cone) is joined by straight lines to a conic. The cylinder is a degenerate cone and so is also a quadric surface. Cones (and hence cylinders) may be circular, elliptic, parabolic or hyperbolic. Such cones are known as cones of the second degree; it is possible to create more complex cones by joining other curves to the vertex of the cone and so create cones which are not quadric surfaces since their slices would no longer give conics.

These surfaces are often made as surfaces of revolution, as described in chapter 5. A spheroid is shown there as figure 5-4, a circular paraboloid as figure 5-3 and the two hyperboloids as figure 5-5.

The original Henrici and Brill versions of Sliceforms in the photographs at the beginning of the book are all quadric surfaces.

Quadric surfaces and circular slices

At the beginning of the chapter, we saw how any slice through a sphere is a circle. It is also possible to build other quadric surfaces using slices which are circles. In fact this was origin of Sliceforms at the end of the nineteenth century. It is also the way that most people know Sliceforms since this method is also described in the classic book "Mathematical Models" by Cundy and Rollett and how I came to make my first models.

A general method for constructing with circles

The following method is a general one for constructing surfaces using circles as slices. It does not just apply to quadric surfaces, and indeed can be developed to create new surfaces and taken on a stage to use other slices of predetermined shape. This is covered in detail in chapter 9 in the section "Surfaces from circles".

The following example shows how to construct an ellipsoid . Other grids for different conics are discussed following the stages for making this model.

Stage 1 - Draw the conic and add the grid

Figure 2-10 shows an ellipse with a grid added. As with the two sphere examples in figure 2-2, the angle of the grid lines can be varied to give a number of models which behave differently as they open. To create an ellipsoid using circles, you *must* make the two sets of grid lines identical, that is you must use the axes of the conic as axes of symmetry. Why this is so and why you cannot have the two perpendicular axes as the grid axes is described at the end of the construction steps.

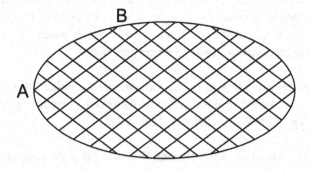

figure 2-10

Stage 2 - Draw the circles and add the slots

Each one of the lines defines a circle, with the length of the line defining the diameter of the circle. The position of the slots is determined from the grid lines in the other direction where these lines intersect the defining circle. As with the description of making the sphere, since there is symmetry, many slices are the same. In this example, there are six different slices, out of a total of eleven slices in each direction. Figure 2-11 shows the construction of the smallest slice (line AB in figure 2-10):

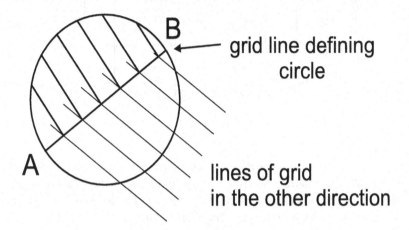

B ←— grid line defining circle

lines of grid in the other direction

figure 2-11

So to construct this slice, copy the length of line AB from line AB in figure 2-10 and draw a circle with that diameter. Mark of the position of the slots where the lines of the grid in the other direction cut AB (either use a pair of compasses or dividers or use the technique using the "ruler" shown in figure 3-4). Draw the slots perpendicular to the defining diameter AB.

Stage 3 - Cut out and assemble

When you have designed each circle, cut out the required number of each design and the slots. Then assemble and consolidate as described in the method for any Sliceform model in chapter 1.

This method applies equally well to making the spheres in figures 2-5 to 2-8. Because of the deformation the spheres are also models of ellipsoids and ellipsoids models of spheres. The only difference, as with the 36° and 90° grid spheres is how their shape changes when deformed away from a sphere and folded flat.

Why must the grid be symmetrical?

You cannot use a grid like the one in figure 2-12, or indeed any one which is not symmetrical to create two sets of circles.

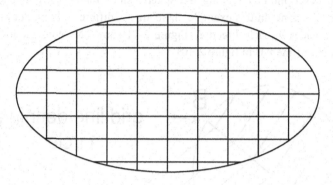

figure 2-12

If you did, the two slices in each direction would be incompatible. This would be equivalent to rotating the ellipse in two directions at once, and only one rotation is needed to define the surface. This may not be obvious if you cannot easily picture what would happen. Consider creating two circular slices which would result from the minor and major axes of the ellipse from the grid of figure 2-12. These are shown in figure 2-13 and you can see that the two circles would not fit together. Rotating the vertical axis to give the smallest circle is obviously not the same as rotating the horizontal axis to give the larger circle. There is only one ellipse that gives two identical slices: when the axes are identical, that is for a circle.

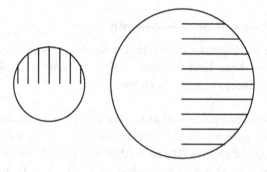

figure 2-13

If you want a spheroid, that is an ellipse rotated about its major or minor axis, then you could use the method described in chapter 5 which describes surfaces of revolution (there is also another method described at the end of this chapter). Suppose you create a spheroid using the grid of figure 2-12, with the vertical lines of the grid as circles (that is you are rotating about the horizontal axis) then the slices in the horizontal direction will be ellipses and not circles. Slicing any ellipsoid other than in two symmetrical directions about an axis as shown in figure 2-10 will give an ellipse. Moreover, it is possible to slice any ellipsoid in this way to give two sets of circles with one exception. If the ellipsoid is a spheroid, then the two sets of circles become the single set perpendicular to the axis of rotation.

Example models of quadric surfaces

The following grids show how you would construct a range of quadric surfaces using circular sections. There is one quadric surface missing from these: the hyperbolic paraboloid. It has no circular sections and is constructed as described in chapter 6, as a ruled surface.

The grid for the paraboloid has a symmetry about an axis through the focus:

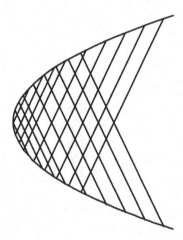

figure 2-14

The model is, of course, only a fraction of the mathematical surface since that is infinite. Consequently, the grid is incomplete and the slices drawn on the right part of the grid would have more slots in them were the model to be made larger. Figure 2-15 shows the parabolic outline with the V-shape of figure 2-14 hidden at the left.

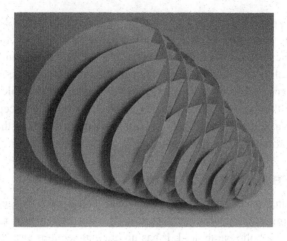

figure 2-15

A hyperbola has two axes of symmetry, giving rise to hyperboloids of one and two sheets as shown in figure 5-. The grids shown in figure 2-16 are for the same generating hyperbola.

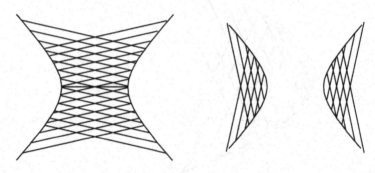

figure 2-16

Note how the one of the right, which generates the hyperboloid of two sheets is actually two separate, identical, objects. The only way it can be see as a complete model is if it is mounted for display. This can be seen in the Brill models photograph at the beginning of the book. The hyperboloid of one sheet behaves in a slightly different way when flattened. As can be seen in figure 2-17, it is very long when open (with 14 slices in each direction) and is still three dimensional in the other direction.

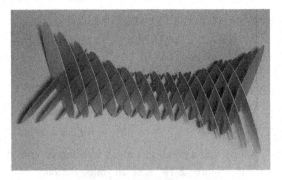

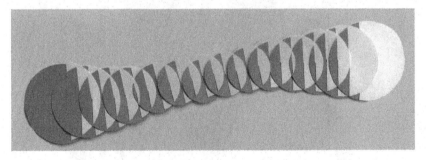

figure 2-17

It is perhaps pertinent to note that making an ellipsoid using the grid from figure 2-10 is not really necessary if you have made a sphere, since the sphere deforms to an ellipsoid.

A cone can also be generated using this method:

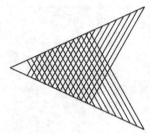

figure 2-18

and, the model (strictly half a cone) has much in common with the paraboloid as can be seen in figure 2-19 and in one flattened form is almost indistinguishable.

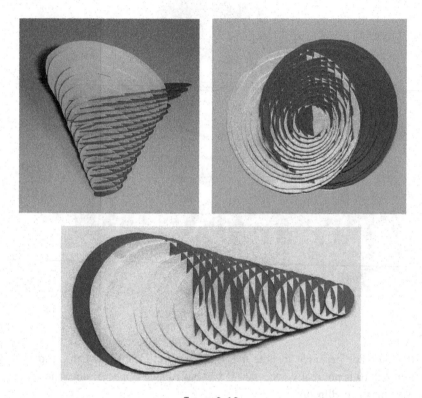

figure 2-19

Since the cylinder is a cone where the vertex has been moved to infinity, it can also be created using a set of circles:

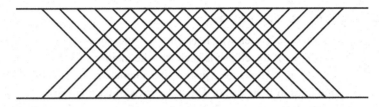

figure 2-20

This means that each slice is now identical and so it is a model which is easily designed. The model has an outline of a pair of parallel lines when folded flat in one direction and is like a three dimensional cylinder when folded flat the other way.

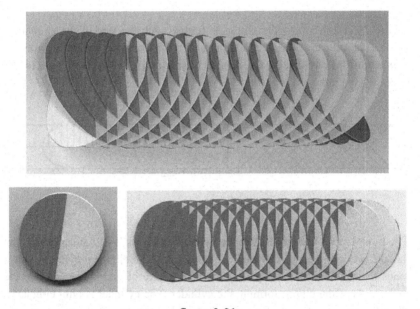

figure 2-21

However, it does not create a circular cylinder, but an elliptical one. It obviously cannot be circular for the simple reason that the circles, as seen by the grid in figure 2-20, are inclined at an angle to the axis of the cylinder. A similar lack of rotational symmetry applies to the other models created using circles. So, for example, the cone in figure 2-19 is not a right circular cone; it has an elliptical base. Moreover, in all these cases, the shape of the ellipse changes as you flex the slices.

Making a circular cylinder

In order to make a cylinder which is circular, the slices have to be made elliptical. The shape of the ellipse is such that the major axis is defined by the angle of inclination of the lines, just as the diameters of the circles were defined by the lines in figure 2-20. The minor axis of each ellipse is the width of the lines of as shown in figure 2-22. If an artist draws a circle in perspective and the circle is not perpendicular to the picture plane, then the image is an ellipse. We have the same situation here in reverse. We want to look along the cylinder and see it as a symmetrical circle. The plane of the slice on which we project the circle is at an angle to our viewpoint and so it becomes an ellipse. The ellipses seen as parallel perspective views of each circle are the same all the way along the cylinder.

Figure 2-22 shows how the ellipse is formed with the size of the equivalent circle. The ellipse is this circle squashed to fit between the lines, with the diameter of the circle being the same as the major axis of the ellipse slices. All slices are the same.

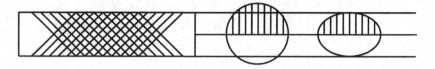

figure 2-22

The model looks very much like the cylinder in figure 2-21 when opened, but has the property that it rolls easily. You can roll the elliptical cylinder, but it tends to reach equilibrium quickly since its centre of gravity is closer to the minor axis of the ellipse forming the cylinder. The cylinder is only circular at the position when the elliptical slices are inclined at the design angle. As it is closed up in either direction it becomes more and more elliptical.

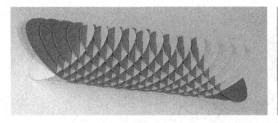

figure 2-23

Making a spheroid

This technique for making a cylinder can also be applied to any other quadric. Figure 2-24 shows how the ellipsoid in figure 2-10 would be used to determine the shape of one of the ellipses. Each line of the grid must be used to create an ellipse in the same way as for the circular cylinder in figure 2-22. The elliptical slice on the right has been drawn for the marked line on the grid. The length of this line is the length of the major diameter of this slice. The minor diameter is the distance between the lines.

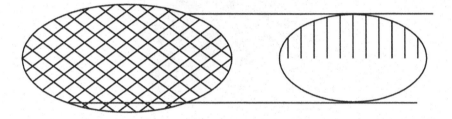

figure 2-24

The spheroid looks very much like the ellipsoid generated by the grid in figure 2-10, and will only be a spheroid if the Sliceform is opened to the same grid angle as was used to create it. In this state, as with the cylinder, it will roll.

figure 2-25

The spheroid in figures 2-24 and 2-25 have an axis of rotation which is along the major axis of the ellipse. Figure 2-26 shows the spheroid where the axis of rotation is along the minor axis of the ellipse and produces a model which is much flatter.

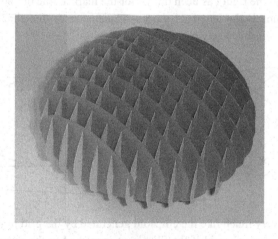

figure 2-26

Other Sliceform quadric surfaces

Quadric surfaces can be sliced in many different ways, and other methods can be used to generate them. Other Sliceform quadric surfaces are described as follows:

- a cone generated as a surface of rotation is shown in chapter 5, Surfaces of revolution"
- a paraboloid generated as a surface of rotation is shown in chapter 5; this has very special properties since the slices can be rearranged to form a

different model as described in the section "Creating new surfaces" in chapter 9.

- various hyperbolic paraboloids generated as a ruled surface are shown in chapter 6, "Ruled surfaces"
- surfaces derived from the hyperbola and/or parabola are infinite, so you can only create a model of part of the surface; you can also extend this to creating models using part of a sphere or ellipsoid and examples of these are described in chapter 9.

Chapter 3
Contours and profiles

This chapter describes how to convert contour maps into models using Sliceform techniques. It also describes how to use profiles (either mathematical or freely determined) to build models.

Contour maps of surfaces

Whereas we see surfaces all around us everyday, some of them are very big and we do not always appreciate that they are surfaces. The obvious case is the earth which is so big that we see it as a flat plane unless we look for subtle clues like the curvature of the horizon. Indeed curvature is a very important part of a surface. Going up or down a hill is another obvious way we know that we do not live on a flat plane. Travelling over a large surface is helped by having a map and most maps are flat. So many maps have surface information on them as contours, and these contours can be used to regenerate a surface in three dimensions. A contour is a line joining points of equal vertical distance relative to a reference height (usually sea level).

The invention of contour maps

Contour lines were first used for mapping river depths and the first known map is by the Dutch Surveyor Pieter Bruinsz in 1584. They were reinvented many times in the next two centuries because communication was not as rapid as it is today. Contour maps were invented in Great Britain to help solve a problem in finding the mass of the earth. These were based on Newton's work on gravitation and the idea that if you hang a plumb line near to a mountain it will be deflected by the mass of the mountain. The English Astronomer Royal Neville Maskeleyne carried out many accurate measurements in 1774-76 in Scotland at various places around a mountain called Schiehallion (or Schehallien) just North of Perth in Scotland. Maskeleyne's method used the deflection of a plumb line against fixed stars as it was pulled by the mass of a mountain. The mountain was found by Charles Mason (of Mason-Dixon line fame) who had been sent by the Royal Society on a tour of Scotland to find a suitable one. He then asked for help in finding the shape of the mountain from the mathematician Charles Hutton who made a flat map. In trying to make sense of the data, he joined points of equal height and realised that this showed him the shape of the mountain as a contour

map. Figure 3-1 shows part of Hutton's map. You can see his contour line marked
as a series of straight dotted lines around the central peak. Note how the peaks are
marked on the map with a precursor of contours, as a shading which shows their
steepness.

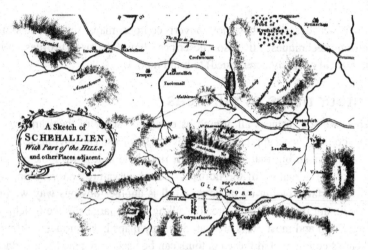

figure 3-1

Properties of contour maps

Sets of contours show lines on a map normally representing lines of equally
spaced heights. The relative spacing of the contours at a particular place then gives
an indication of the slope at that place. When contours are close together, this
means that the slope of the terrain is steep, and conversely where there are few
contours that it is flat. Hutton's map in figure 3-1 shows only the impression
of steepness. Other features that show up are ridges and valleys and the peaks
where concentric contours no longer have any other ones within them. The
map in figure 3-3 shows a hill with two peaks and a valley between them. The
peak on the right is not as high as the hill on the left.

Contour maps give you a good idea where mountains and valleys are, but it needs
experience to interpret them fully and they are not always easy to visualise in three
dimensions. As well as models helping in this respect, they are often required for
architectural or geographical purposes.

Types of contour maps

Contour maps may be drawn to show other equal properties besides geographic terrain. They may not map surface height, but they may show other properties like barometric pressure on weather maps or magnetic dip.

Mathematical equations can also be visualised by means of contour maps to show algebraic surfaces. In making a model, its is usually better to use the equation as described in chapter 7 since you can usually produce the model faster. Figure 3-2 shows a contour diagram of the monkey saddle surface described in chapter 7 whose Sliceform model is shown in figure 7-37. The contour has been shaded according to height in order that there can be differentiation between a mountain or valley slope. Thus three of the six sectors are mountains and three valleys as will be apparent from looking at the three dimensional representation in chapter 7.

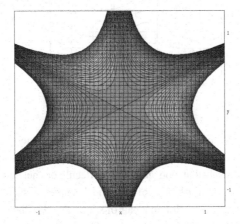

figure 3-2

Most of models created in chapter 7, "Surfaces from equations - Algebraic surfaces", involve explicit equations, where the values of cartesian coordinates can be plotted from the equation. However, some equations can be expressed implicitly (none of its variables is isolated) in an equation so that the equation cannot be solved. Using numerical calculation, it is possible to plot such equations, but then further work is required to create the slices for the model. In some cases this calculation generates sets of curves which can be seen as a contour map which can then be used to define a surface. Making a Sliceform model is then an ideal way to demonstrate the surface. Examples of these are shown later in this chapter in the section "Mathematical Contours".

Turning a contour map into a Sliceform model

The first step in creating the Sliceform model is to create a set of profiles and use these as slices. By creating profiles in different directions you can then slot them together and the Sliceform model of the surface enables you to convert a two-dimensional map into a tangible solid.

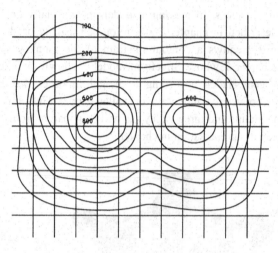

figure 3-3

First mark a grid on top of the contour map. As described in chapter 1, make the lines of the grid open ended, that is do not create the profile at the ends of the grid. The grid lines should be such that you get enough detail for the map, without having to do too much work. Each line of the grid becomes a slice and each intersection of grid lines becomes a slot on a pair of slices. Position the grid so that you get as many grid lines on the peaks as possible. This is not always feasible, as in the example in the diagram. The right-hand peak highest contour falls within a grid square and will not be apparent in the model since it reaches its apex over a hole in the grid. The peak in the model appears slightly away from the actual peak. This is like the discussion of making sure you see maxima and minima in geometrically defined Sliceforms as described at the end of chapter 1.

Next move along each line of the grid creating a "ruler" showing the positions of the heights. Use the edge of a piece of paper placed along the grid line and make marks on this paper to define where the contour lines intersect with the line. The ruler marks are shown in figure 3-4. When you have made the marks, write a height contour identifier next to the marks to distinguish them and indicate where peaks and troughs occur.

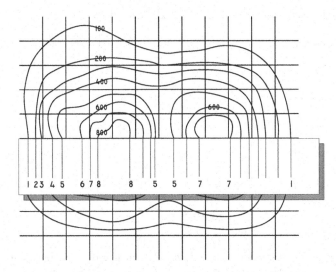

figure 3-4

The next step is to turn these marks into heights of the slice. You could work with the same edge of the paper, or you could copy the marks to a line on another sheet of paper. It may be convenient to work on a sheet of graph paper, which will help with giving perpendicular lines. Whichever way you decide to work, create the profile by raising lines above the "ruler" marks in proportion to the height of the contour to which they correspond. Make sure you keep the same proportion as the scale of the map; otherwise, the heights will be distorted compared with distances on the ground. Mark the lines faintly since they are only construction lines.

Draw in the profile by freehand sketching. Where there is a gap at a peak or trough in the profile, you should guess the shape. This is not too difficult if you follow the slope of the lines either side of the gap. You do not have any information in the map apart from the spacing of the contour lines to guide you anyway. Similarly, extend the shape of the slice to the edge from the outermost know contour. Also add a base to each slice by adding a rectangle below the base of the height lines with sufficient depth that the remaining card above or below the smallest slot on the model is substantial enough for the slice not to tear. Finally, mark the slot lines to correspond to the grid lines. Take care to mark grid lines and not use the height lines from the ruler. Figure 3-5 shows a slice built from the ruler used in figure 3-4.

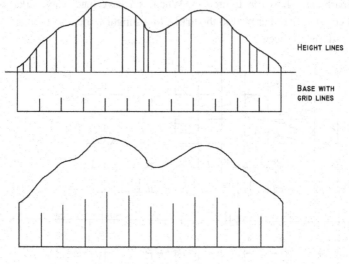

figure 3-5

The Sliceform model of the contour of figure 3-3 is then shown in figure 3-6. Folding such a model flat has the benefit of easy storage. For a more realistic, non-deformable model, you can use the plaster sculpting technique described in the section "Solid Models" in chapter 9 to give a solid model. Using this technique, you can smooth the surface completely, filling in areas which are the holes in the Sliceform model.

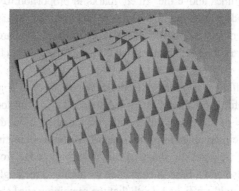

figure 3-6

Mathematical contours

The following descriptions of models should be looked at in conjunction with chapter 7 which uses explicit mathematical equations to define the surfaces and plot the slices. The models here are ones which are not always seen as surfaces, and in some instances are very much on the side of mathematical art rather than pure mathematics.

Families of curves

Mathematics is about patterns and relationships. Often these are expressed through the strict mathematical language of equations and the formulation of such concepts as rotations and translations. A specific curve is defined by the constraints, but if these are varied then a family of curves is defined. This is best seen by looking at the examples in the next few pages.

Families of ellipses

A common way to draw an ellipse is to use the property that the sum of the distances of any point on the ellipse to the two foci is constant. Tie the ends of a piece of string to two pins at the foci and put pencil in the loop. Then draw the ellipse by keeping the string taut.

figure 3- 7

Now, if you keep the same foci and vary the length of the string, the family of confocal ellipses looks very much like a contour map and can be treated as one for making Sliceform models. The result (figure 3-8) shows that the line between the foci (which results when the string length is equal to the distance between the foci) gives a sharp ridge in the contour.

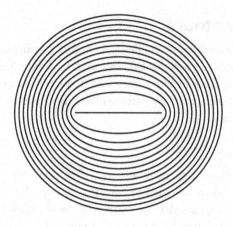

figure 3-8

This model could be created from the contour, but it can also be created from by plotting as an equation using the methods described in chapter 7. Such methods allow the surface to be seen in the computer before modelling. This model is not particularly interesting and hard to make since the contours show a very steep surface, like a chisel with the central line being its cutting edge. A variation on the equation for confocal ellipses enables the analogous confocal hyperbolae to be drawn also. There is a more complicated string method which can be used to draw hyperbolae, like that used in figure 3-7 for the ellipses, and this yields a more interesting surface. Details of these are given in chapter 7.

Cartesian ovals

The Scottish mathematician and physicist James Clerk Maxwell is famous for the set of equations bearing his name which describe electromagnetic waves. His collected papers fill a number of thick volumes and show many related to the study of pure geometry. His first paper was "On the Description of Oval curves and those having a plurality of foci" which was communicated to the Royal Society of Edinburgh in 1846 by Professor Forbes, because he was only fifteen.

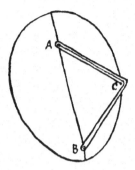

figure 3-9

Maxwell generalised the construction of the ellipse by winding the string around the foci in different ways before drawing the curve. In figure 3-9, one end of the string is tied to focus A and the string is then taken around focus B and then back around focus A before the pencil is attached to the other end at C. The curve is drawn by keeping the string tight. Even though there is a constant length of string, sum of the distances of a point from the foci is no longer a constant. The sum of three times the distance to A and twice the distance to B is a constant (the length of the string). Winding in different ways can produce a whole set of curves. If there are m strands of string to one focus and n to the other, an ellipse results when m and n are both the same.

The curves produced by Maxwell's methods are ovals which have only one axis of symmetry, unlike the ellipse which has two. In fact they are a type of curve called a Cartesian oval. These were discovered by the French mathematician and philosopher Rene Descartes in 1637. The shape of the curve varies according to the ratio *m:n* but also with the ratio of the length of string relative to the distance between the foci. Sometimes the curve is in one part, sometimes it splits into two parts one inside the other. It may be egg shaped or nearly circular. Figure 3-10 shows a set of curves of the family in a form which suggests a contour, then with the component curves separated so that you can see which parts are paired or singular.

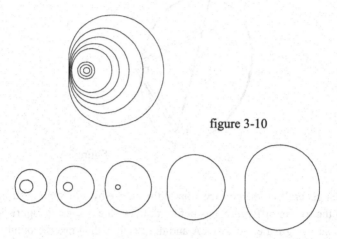

figure 3-10

Maxwell extended the principle to more than two foci. Figure 3-11 shows how he used three foci; the dotted lines at the top for obtaining point D′ indicate how he had to rewind the string. He does not appear to have thought about surfaces in this way, possibly because it is not easy to do mechanically. But making Sliceform surfaces using this method is possible with a computer, and is discussed in chapters 7 and 10.

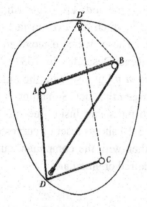

figure 3-11

Maxwell wrote more about the subject later in his life and figure 3-11 shows an oval which is from a later paper where he also considered families of these curves produced by varying the length of the string. A computer simulation of changing the string length is shown in figure 3-12.

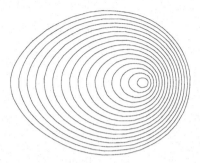

figure 3-12

Note how the "peak" (considering the family of curves as a contour) is asymmetrically placed. This suggests a set of contours even more than the curves in figure 3-10 and offer many possibilities for exploration. A Sliceform model of a surface from figure 3-12 is shown in figure 3-13.

figure 3-13

His extension to the cases where there are more than two foci gives even more scope for creative activity, since families of curves can be obtained in many more ways that with two foci; as well as the ratio of the distance to three foci offering more possibilities, the foci can be the points of different triangles. These curves are easy to draw mechanically using the methods described here, but plotting them as equations is not as easy.

Figures 3-14 and 3-15 show two families of curves obtained from three foci. In both cases the foci are placed at the corners of an equilateral triangle. The difference is that the sum of the distances to the three foci is equal in figure 3-14, but in figure 3-15 the constant is the sum of twice the distance to one focus and the distance to the other two foci, which is equivalent to saying that there are two strands of string to two of the foci and two strands to the other one.

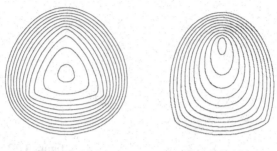

figure 3-14   figure 3-15

Note how figure 3-15 shows the asymmetry because of the top point being the one where the distance is multiplied by 2 and so attracts the curves to itself. Figure 3-16 shows the Sliceform model that results from using the contour of figure 3-14 to define the surface.

figure 3-16

Such methods can obviously be extended to define surfaces in three dimensions if there are at least four foci which are not all in the same plane. Methods for plotting the curves on the computer and for creating slices of such surfaces are described in chapter 10.

Another way to create ovals

This method is commonly found in books of mathematical recreations. The question asked is "Can you draw an egg shape using a pair of compasses?" and like many of these recreations is a trick method. You solve it by wrapping some paper around a tube and then drawing with the compasses in the normal way using one sweep of the compasses.

figure 3-17

The ovals so produced are symmetrical about two axes and you can get a good egg shape by drawing half of the shape when the paper is flat.

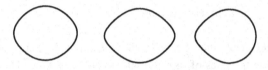

figure 3-18

The family of curves obtained by varying the opening of the compasses shown in figure 3-19 gives rise to the Sliceform model shown in figure 3-20.

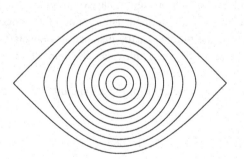

figure 3-19

figure 3-20

This technique offers much scope for exploration. All of the ellipse and Maxwell methods using a piece of string could be explored to create a wide range of new families of curves. The tube could also be replaced by other shapes, for example a cone. The surfaces would need to be ones onto which a piece of paper could be placed without distortion, so a sphere is not suitable.

Other possibilities for mathematical contours

Many curves can be plotted as a series of contours. The following are suggestions for other curves to explore:

- the magnetic curve and its corresponding equipotential curve, which can be visualised as the lines of force around a bar magnet (see David Wells {1} figure 7.25);
- parallel curves, where a curve is drawn a distance from the defining curve; this is not the same as simply magnifying a curve since the parallel curve is constructed by measuring a fixed distance along the normal;
- the Cartesian oval and other relationships described by Maxwell form a group of curves which are defined using bipolar co-ordinates; whereas the ellipse and Cartesian ovals were defined using sums of distances, other functions can be used; if the product of the distance is a constant for two foci then the result is a set of Cassininian ovals, which are described using equations, in chapter 7;
- families can be derived from tripolar co-ordinates and other multi-polar co-ordinates.

References for exploring these curves are Lockwood and Yates.

Models from profiles

You can also build Sliceform models from sets of profiles. There are a number of ways to obtain profiles:

- profiles obtained from objects such as sculptures, everyday objects or even your body which allow you to make a Sliceform model to match the original three dimensional object; the most difficult problems in constructing models with three dimensional objects are ensuring even spacing of the slices and creating an accurate profile of each slice.
- using a profile which is repeated in space, that is extruding the profile;
- as with contours, there are many mathematical curves which are ideal profiles, and the one discussed in this chapter illustrates how some of the contour methods can be used with these when it may be possible to plot an equation in two dimensions, but it may not be easy to obtain the slices in the perpendicular direction.

When slices are only available in one direction

When you have a set of profiles in one direction, it may not be possible to get an accurate set of profiles in another direction. It is possible to make a model which does not slot together (which could apply to all models, in fact) and hence does not fold. An example is shown in figure 3-21. Alternatively, or the slices in the other direction can be constructed from your knowledge of the first set, as shown in figure 3-22.

Making a non-folding model

The slices in each direction of a model support one another. There are a number of ways to make a slotted support to make a model that does not fold. The most basic is to cut a set of thin grooves in a piece of wood or even expanded polystyrene into which the slices fit. Another way is to build a base with slits in it from a piece of folded card and make a model like the one in figure 3-21. When cutting the slits in this card, just make one cut, do not cut slots as for a Sliceform. The slit then acts as a pincer mechanism and provides extra support by the way it grips the slice. No slots are required in the slices in any of these methods.

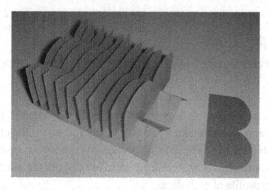

figure 3-21

The model (the cycloid surface) in figure 3-21 is also shown in figure 3-27 as a normal Sliceform model, to demonstrate how the same result can be used for different effect. The description of the mathematical construction of this profile obtained by plotting a series of curves is given later in this chapter.

Slices in the "other" direction

If you have a set of slices and the slots are drawn on them then you know the height of the slice at one particular place on the model in both directions. The spacing in the "other" direction is arbitrary, but looks best if the slots are the same spacing as on the original profiles. You can thus build up a set of heights for a particular slice in the "other" direction and free hand draw the curve at the top of the slice. Figure 3-22 shows two slices of the cycloid surface of figure 3-21 obtained by plotting a series of eleven curves which become the x-direction slices, together with a y-direction slice to show how they were converted into the normal Sliceform model shown in figure 3-27.

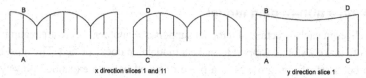

x direction slices 1 and 11   y direction slice 1

figure 3-22

Line AB of x direction slice 1 becomes the first line in y direction slice 1. Similarly, line CD of the last slice in the x direction (slice 11) becomes the last slot line in the y direction slice 1. Other slots have been marked in. You need the height first in order to determine the shape of the top of the y direction slice. In *Designing with a square grid* in chapter 1 (see figure 1-5), it was show how the outside of the grid does not form a slice because it would not have a support. So all slices overlap the slices in the other direction. When creating slices in the "other" direction as in figure 3-22, remember to include the overlap part of the slice in the y direction that has no constructed slice in the x direction so that the ends are formed for the y direction slice.

This method is particularly easy when you are making an "extrusion" profile model as described later in this chapter. All the constructed slices in the "other" direction are then rectangles.

Using a sliding pin profiler

When cutting wallpaper, you often need to copy the profile of a moulding (for example a skirting board) to be able to cut the shape in the paper. Do-it-yourself shops sell a tool for this which is a rod with a series of pins which slide perpendicular to the rod. This method is ideal for making one profile, but is much harder to work with when trying to create a series of profiles from an object or sculpture, because there is no automatic creation of a spacing for the slices. One way to overcome this is to mark a series of lines on your object using a felt tip pen and use these as a guide defining them from positions of regularly spaced pins on the profiler tool held in the perpendicular direction to the way you are working.

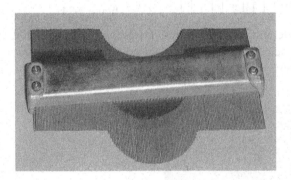

figure 3-23

Once you have a set of profile slices in one direction, creating a set in the other direction is more accurately achieved using the method described in "Slices in the "other" direction" above.

Mathematical profiles

It is very easy to plot curves and turn them into Sliceform models. The following two examples show the two basic methods of doing this.

The "extrusion" method

The idea used here is to take a curve and plot it to form all the slices in one direction. This means you only design one slice and repeat it. The slices in the other direction are then all rectangles whose height is defined by the height of the curve at the position of the slices. The spacing of the slots on the rectangle is arbitrary, but looks best if the slots are the same spacing as on the original profile. The number of slots you choose on the rectangle also decides its size.

The model in figure 3-25 is made from eight slices in the shape of a sine wave, with slices as shown in figure 3-24. Note that it is important to make sure that the slots (and hence rectangular slices) are positioned on maxima and minima as indicated with the slots extended by dotted lines.

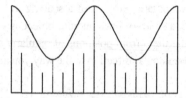

figure 3-24

The assembled model in figure 3-25 shows interesting effects when the slices are not quite perpendicular and when it is folded flat the patterns are unlike most other Sliceforms.

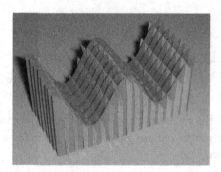

figure 3-25

The cycloid surface

The profile method can be used to create a host of new surfaces if you have an equation with a parameter that can be varied. Varying the parameter defines a set of the slices of the surface. This model is an example of creating a surface from a sequence of equations where it is not easy to determine the "other" slices.

Consider a wheel rolling along a line. The path a point on the wheel traces a curve as the wheel rotates which is known as a cycloid. If the point is on the rim of the wheel (top of figure 3-26), then the cycloid is in the form of an arch. The cycloid in this form has been studied by many famous mathematicians, and was probably first conceived by Mersenne and Galileo in 1599. Sir Christopher Wren proved that the length of one arch is 8 times the radius of the rolling circle in 1658. Roberval proved that the area of one arch is 3 times the area of the rolling circle in 1634.

The curve changes shape as the point moves to and from the centre. If the point is inside the wheel, then the curve is known as a curtate cycloid (centre of figure 3-26) and if it is outside the wheel's disk then it is a prolate cycloid, (bottom of figure 3-26).

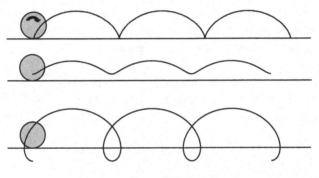

figure 3-26

The prolate cycloid, because the curve intersects itself, is not suitable for the model, but varying the position of the point across a diameter, yields a set of curves which give an interesting surface.

If a point is at a distance b from the centre of the rolling circle which has radius a then the curve can be plotted using the parametric equations:

$$x = a\theta - b\sin\theta$$
$$y = a - b\cos\theta$$

The curve is infinite if the wheel just goes on rolling, but the Sliceform model is based on only two revolutions. To obtain the slices, the curve is plotted eleven times with the parameter b taking values that vary from of $-a$ to $+a$, where a is the radius of the circle. This means that the point on the wheel starts either on the ground or as far as possible from the ground at the other end of the diameter. Consequently, the two outside slices have curves which are shifted with respect to one another. (These are the two slices to the left in figure 3-22.) The central slice of the model is a rectangle since the point on the wheel lies at the centre and so does not change its height above the ground as the wheel rotates. As with the models described above, add a base to each slice by adding a rectangle below the base of the height lines.

The model with only the slices drawn from the curves is shown with a support in figure 3-21. With the slices added in the other direction so that the model can collapse gives figure 3-27.

figure 3-27

Chapter 4
Properties and types of Surfaces

This chapter is not about constructing particular Sliceform models, but describes types of surfaces and their properties as a background to surfaces described in other chapters.

What is meant by a surface?

The Concise Oxford English Dictionary defines a surface as:

The outside of a material body;
any of the limits terminating a solid;
Geom, a set of points that has length and breadth but no thickness.

to which can be added that surfaces exist in space and are the equivalent of curves in the plane.

Surfaces are all around us and we generally do not give a thought to their types or their properties. The French mathematician Gaspard Monge (1746-1818) who is credited with the invention of descriptive geometry, often called engineering or technical drawing now, gave a description of a surface which is pertinent to creating surfaces for Sliceform models:

Every Curved surface may be regarded as being generated by the movement of a line, either constant in form while it changes its position or variable at the same time both in form and position.

As the history of the study of surfaces gathered momentum throughout the nineteenth century, other methods evolved to describe surfaces using different approaches. Some surfaces are known because they were the solutions to a particular problem (for example the surface formed as a set of points equidistant from a point, namely a sphere) or were arrived at by the study of an equation or a transformation (such as Steiner's Roman surface, figure 3-3). Although there are books which describe curves, such as Lockwood, Shikin and Yates, there is nothing as comprehensive for surfaces. The classic is Gerd Fischer's book "Mathematical Models" which has photographs of many

classical surfaces and a commentary on the mathematics. In either case there
are many classifications of both curves and surfaces, so it makes sense to look
at surfaces from the point of view that we want to make a Sliceform model as
opposed to any other.

What makes a surface suitable as a Sliceform?

This is not an easy question to answer, and in some cases it is easier to ask the
question negatively since there are properties of surfaces which make them
unsuitable for different reasons. It may also be possible to choose a different way
of slicing to overcome some of these problems. Often the method of generating
the surface is adapted to making the Sliceform.

An important point to bear in mind when creating a Sliceform model is that the
model needs "substance". This means that there needs to be a way of creating a
complete slice not just its edge. For a closed surface like a sphere the inside of the
sphere can be used. However, to create a Sliceform of an open surface, such as a
flat plane in its simplest case, it is necessary to invent the rest of the slice to add to
the edge which is the surface. This concept of the edge being the surface is often
taken for granted and is usually not an issue when creating a model. In chapter 9,
"Variations and explorations" when discussing ways to develop models, part of a
spherical surface is shown as a bowl (figure 9-10). This means more than one
model can be made from the same surface.

Unsuitable properties

Because a surface is modelled by a set of slicing planes, the commonest reasons
why a surface is unsuitable for modelling is that the slices are not easy to make or
that they separate into a number of pieces.

Whereas it is possible to make a model of part of infinite surfaces such as the
functions described in chapter 7, by building up the substance of the slice, there
are some instances where this is not possible. Such a case is the helix shown in
figure 6-3 in chapter 6, "Ruled surfaces". Extending the surface to make slices
would destroy the ability to see the surface. It would be possible to make a model
where a helical structure is built on a cylinder, for example. While this is not the
surface you might have set out to make, it shows how sometimes it is necessary to
adapt the construction of the Sliceform in order to make the surface.

Some surfaces present this problem when they intersect themselves in a complex
shape. An example of such a surface is Kuen's surface shown in figure 4-1.

figure 4-1

Other unsuitable surfaces have lots of holes and singularities. Typical of these are Kummer's surfaces like figure 4-2 (see Fischer for more details) which have tetrahedral symmetry. Slices become separate parts and the model cannot hold together.

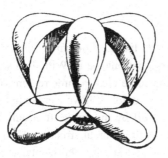

figure 4-2

Steiner's Roman Surface (which is like a tetrahedron where each face has been pushed to the centre) is another surface with singularities which are three lines and holds out even less hope of being able to provide suitable positions for slicing. This also has tetrahedral symmetry and is shown in two views in figure 4-3.

A different reason is that it is difficult to plot the surface because it is described by a mathematical function which is implicit rather than explicit. Chapter 7, "Surfaces from equations - Algebraic surfaces", and chapter 10, "Using the computer", describe methods to plot such problem equations and determine the shape of the slices.

Sometimes it can be time consuming to find the best position to slice a surface if you do not know what it looks like, even though you know a logical way to design it. For example, the surface formed as a set of points where the sum of the distance from a number of points is space is a constant. This is an easy surface to specify but difficult to plot. I have been unable to find a reference to such a surface in the literature. However, this is a case where the Sliceforms method is ideal for creating such a surface and a computer method is described in chapter 10, "Using the computer".

A general problem with creating surfaces as compared with drawing a curve in the plane is that we live in space not outside it. With a closed curve like the limaçon shown in figure 4-4, the curve creates three regions of the plane. The infinite outer region is the one we would see if we were to live *in* the plane and were not inside the areas fenced off by the curve. It is only because we can look from outside that we can see the internal structure. If we extrapolate the curve to a solid object, by rotating it along a horizontal axis symmetrically through the double point (the point where the curve crosses itself), then the internal parts are invisible to us as viewers in three dimensional space. In creating this solid of revolution, the shaded area would be non-solid, that is a hole. Looking from outside we could not see this. One way create it as a Sliceform, would be to make it as two halves, so that the internal structure could be seen.

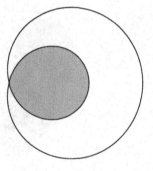

figure 4-4

There are another set of surfaces which are also unsuitable and too complex to describe in the context of Sliceforms. These are the so-called minimal surfaces and are described in Fischer and Peterson.

The Sliceform view surface types and slices

Since we are creating surfaces as a set of slices, it seems appropriate to look at surfaces from the point of view of how do you create them as a Sliceform model. None of the methods are any better than another, since they generate different types of surface, although some may be easier to construct for individual cases. On a personal level, it depends whether you find it easier to work geometrically or algebraically. The following methods are a broad categorisation.

Slice a known surface

The obvious way to make a model is to take a surface you know how to create and slice it up; examples of this are:

- the sphere and the other quadrics described in chapters 1 and 2
- surfaces of revolution described in chapter 5
- the contour surfaces in chapter 3
- though not surfaces in the strict geometric sense, polyhedra are another obvious case; see chapter 8.

The shape of the slices depends on the symmetry of the surface being sliced and the direction in which it is sliced. There may be one or more preferred directions. The classical illustration of slicing a cone leads to the conic sections:

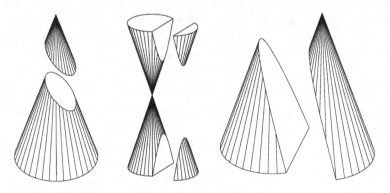

figure 4-5

With a torus a parallel set of slices yield an interesting set of curves as shown in figure 4-6.

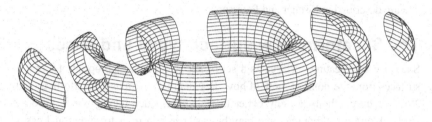

figure 4-6

The curves on these slices are in fact a series of curves known as the spiric sections of Perseus. Because a torus is a surface of revolution of a circle which lies on a plane through the axis of revolution, it is obvious that a torus can be sliced in to give two sets of circles. Firstly, a slice using a plane through the axis, for example the central slice of figure 4-6, is a pair of circles (the ones being rotated). Another set is obtained by slicing through a plane perpendicular to the axis which gives a circle as a rotation of each point of the defining circle. There are also other circles called the Villarceau circles which are described in the section "Slicing the torus" in chapter 7. A torus with non circular sections is shown in figure 9-6, "Variations and explorations" where the defining circle is not on a plane through the axis.

Plot slices of a surface as a function

Where the equation of a surface is known as an explicit mathematical function, this is a very easy way to create slices as described in chapter 7, "Surfaces from equations - Algebraic surfaces". However, if the surface is described by a mathematical function which is implicit rather than explicit, this may not be easy. See chapter 7, for a description of such equations and chapter 10, "Using the computer" for ways to overcome plotting such problem equations.

Generate from a set of slices

This method using the slices as the originator of the surface is described in various places:

- some of the surfaces described in chapter 6, "Ruled Surfaces", fall into this class
- many methods described in chapter 9, "Variations and Explorations", such as creating a surface from sets of circles
- surfaces from profiles are shown in chapter 3, "Contours and profiles"

- surfaces of revolution allow easy definition of slices as described in chapter 5, "Surfaces of Revolution".

Derived surfaces

This heading encompasses a number of miscellaneous surfaces described in chapter 9, "Variations and Explorations":

- creation of one surface from another may be easier when a Sliceform has already been created, for example using the mathematical transform known as inversion to change the shape of the slices;
- a surface may be modified by turning a concavity into a convexity or vice versa and part of a model may be turned into a completely different one
- surfaces can be combined in the same model
- the slices may be assembled in a different way which causes the Sliceform to behave different as it is viewed or flattened, even though the same mathematical surface is being modelled
- two dimensional drawings can be converted into new surfaces

In addition to all of these, it is possible to slice not only in different directions, but so that the slots are not parallel, yielding totally different effects. A model can be made solid by filling it with plaster. These methods are also described in chapter 9.

Relevant properties of surfaces

When you look around you at the surfaces, both natural and man made, that you come across in everyday life, you can see that the study of surfaces could fill many books. Apart from the study of their shape, there is a vast literature on their engineering properties. Both mathematicians and biologists are interested in surfaces which have "optimal" form so that surface forces or areas are minimised. The topology of surfaces, which deals with properties of surfaces which are unchanged after distortions like stretching and twisting, is totally different area. Many of the books in the list of references contain detailed and interesting explanations on the properties of surfaces. The following are a few that relate to Sliceform models.

Curvature of surfaces

The curvature of a surface can dictate the way a surface is sliced. The curvature also helps to define the extent of a surface, that is whether it is closed or whether it is infinite. Curvature of surfaces was first studied by Gauss. He used slices through a surface in different directions to understand how the curvature of the resulting plane curves could be used to measure curvature of the surface. This was published in his *Disquisitiones generales circa superficies curvas* (General investigations of curved surfaces) in 1827.

If you are trying to measure the curvature at a point on a surface, then you could choose to cut a slice through the surface in any direction. To make the measurement, you first find the tangent plane at the point and then slice using a plane which is perpendicular to this tangent plane. When you cut the surface with this plane, the surface produces a curve on the plane. In some cases (like a sphere) the curvature will remain the same if you rotate this slicing plane. Generally this will not be the case and there will be differing curvatures in different directions. If you try and fit circles to the curves, then the radius of the circle is a measure of how much the surface is curving. A small radius means very curved, a larger radius means less curved and a straight line is zero curvature. A surface can have a mixture of curvatures in different directions and the centre of the circle which measures the degree of curvature can swap from one side of the tangent plane to the other as the plane described above, rotates.

If you consider all the curvatures which can occur as you rotate the slicing plane through the point about an axis perpendicular to the tangent plane, then there are always two extreme directions where the curvature of the arc of intersection drawn on this plane reaches a maximum or minimum. These are called the principal curvatures at that point on the surface. If the surface is not plane you can have three distinct situations arising:

- If the centres of the principal curvatures both lie on the same side of the tangent plane, you have either a bump or a bowl surface at that point, also called an elliptic curvature.
- If one principal curvature is zero, then you have a ridge also called a cylindrical curvature.
- If one centre of principal curvature lies on one side of the tangent plane and one lies on the other side, then you have a saddle shaped point, also called a point of hyperbolic curvature. In the case of a saddle the surface will lie on both sides of the tangent plane near the point of interest. For the other two cases the surface is entirely on one side of the tangent plane. (Try to visualise a plane resting at the very centre of the saddle shown at the right in figure 4-7.)

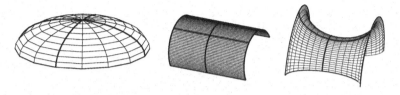

figure 4-7

Surfaces can have mixtures of these types of curvature as you move over them. Any point on an egg or sphere is a bump. The surface between the knuckles on your hand is a saddle shape as indeed is a saddle you would put on a horse. Points on the surface of a cylinder have ridge curvature.

The best type of curvature for a Sliceform model is to have bump curvature all over. The surface is "closed" because the bumps stop it being infinite like a landscape, and all parts of the surface can form the edges of the slices. When you have bowl shaped surfaces, saddles or ridges then you need to be careful where you make the slices. You can end up with a slice which has more than one part.

A normal saddle like the one shown above is convenient for a human to sit on. However, if a monkey sits on a saddle, it needs an extra indentation for its tail. The monkey saddle is a famous surface described in chapter 7, *Surfaces from equations - Algebraic surfaces* and shown in figure 7-1.

As mentioned in the Preface, a number of German model makers produced examples commercially at the end of the nineteenth century, including models to show surface curvature, which do not flatten like Sliceforms. See figure 9-53 for a picture of some of these.

Chapter 5
Surfaces of revolution

This chapter describes methods for creating Sliceform models as surfaces of revolution.

General Description

One common way to generate surfaces is to rotate a line or curve about an axis. Mathematical books on surfaces do not describe many that are generated this way even though the method is simple. It was, however, a popular recreational amusement in the nineteenth century, with toys like the one shown in figure 1.

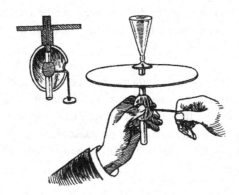

figure 5-1

Various objects could be placed on the holder to give different surfaces. In the one shown, a line produces an inverted cone. A similar effect can be obtained by rolling a piece of bent wire (such as an opened paper clip) between your fingers.

Other simple surfaces of revolution (shown in figure 2) are the sphere, formed by rotating a circle about its diameter, and the cylinder by rotating a line about an axis parallel to the line. Another common one is the torus (the shape of a ring doughnut) which arises when a circle is rotated about an axis parallel to a diameter but outside the circle.

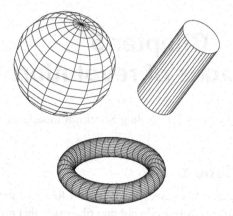

figure 5-2

The conics (the parabola, ellipse and hyperbola) can also be used to create surfaces of revolution. To obtain the surfaces in chapter 2, "Quadric surfaces", the curves must be rotated about their axis of symmetry. However, many different Sliceforms which are not quadric surfaces can be created using a different axis as discussed on in the section "Rotating about another axis" later in this chapter.

The parabola only has one axis of symmetry, but the ellipse and hyperbola have two. The paraboloid of revolution (figure 5-3) is important as a mirror since all light rays parallel to the axis impinging on the surface come to a sharp focus. Placing a light at the focus of the parabola causes a parallel beam of light to be produced, and this is used to create a parallel beam of light in a searchlight or car headlamp.

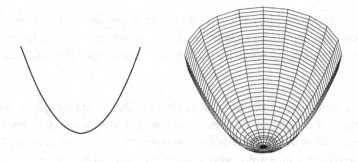

figure 5-3

The two solids generated by rotating an ellipse about its major and minor axes are essentially the same, almost like a squashed sphere with different amounts of flattening according to whether you use the long (major) axis or short (minor) axis for rotation. The one in figure 5-4 is obtained by rotating the ellipse about its minor axis. Ellipsoid-like shapes are found in nature, for example sea urchins or pebbles washed up on a beach.

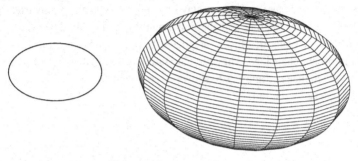

figure 5-4

The hyperbola on the other hand gives two distinct surfaces. One is produced by rotation about the line joining its two foci is in two pieces. Rotation about the perpendicular bisector of this line gives a single surface. They are thus known as the hyperboloids of two and one sheets respectively (see figure 5-5).

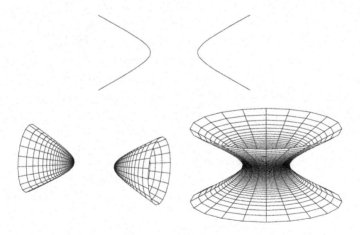

figure 5-5

The one sheet hyperboloid is like the shape seen as cooling towers at power stations. The two sheet one can be made as two separate Sliceforms. With the hyperboloid of one sheet, care has to be taken in making the slices since they consist of more than one piece if the slicing plane is not close to the axis of revolution. A model of a hyperboloid of two sheets must be made in two separate pieces. This can be seen in the Brill models shown at the front of the book.

Surfaces of revolution are very common in the art/craft and engineering worlds where a shape can be made by rotation of the material being worked. In ceramics they are easy to form in clay on a potter's wheel. Symmetrical glass vases are made by rotating and shaping a bubble of glass on the end of a pipe. Many different types of objects are made by turning wood on a lathe, from bowls to ornate chair legs. They are also familiar in metalwork, particularly as goblets or chalices. Figure 5-6 shows a computer created chalice obtained from the curve on the left.

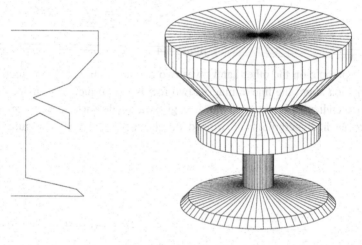

figure 5-6

The Renaissance artist Uccello produced the perspective drawing of a chalice shown in figure 5-7 by rotating a curve. The methods he used are essentially the same as the way such objects are generated today in 3D computer packages.

figure 5-7

Complicated surfaces of revolution like these are not suitable for making as Sliceforms since slices would consist of disjointed parts.

Properties of surfaces of revolution

When such a surface is generated, each point as it moves remains at the same distance from the axis of rotation. This means that the points describe a circle whose centre is on the axis. The circle is a plane which is perpendicular to the axis. These relationships are the most important to bear in mind when drawing or calculating the positions of points of the solid surface generated.

The easiest way to create Sliceforms from surfaces of revolution is to make the slices parallel to the axis of rotation, in two directions at right angles. Looking down on the grid of the slices, with the axis pointing towards you, the surface of revolution looks like a series of concentric circles for each point being rotated. The slices are not circular and do not even look like the curve being rotated, unless they pass through the axis of rotation.

If you centre the grid on the axis of rotation, the symmetry means that there several slices with the same shape. Also each slice has an axis of symmetry about the central line of the grid. When you create the model, you only need to produce one drawing for four slices (or two for the case where the slices go through the axis). This drawing (shown in figure 5-8) looking down on a grid with the circles marked, has the grid lines for four identical slices marked with arrows.

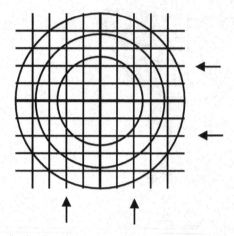

figure 5-8

Although a number of slices have the same shape, the slots in each are different depending on which direction they are on the grid.

The following geometric description of making a cone as a surface of revolution shows how to use these circles to build up the slices and is applicable to other curves. Other methods are described in chapter 7 in the section Curves to surfaces of revolution on page 159, and in chapter 10, *Using a computer*.

Modelling a right cone

This model, shown in figure 5-9 in its open and flattened form, is the easiest one to make as a surface of revolution because it is formed from the revolution of a line as in the Victorian parlour recreation at the beginning of the chapter.

figure 5-9

If you rotate a line which starts on the axis of rotation, the surface generated is a cone with the vertex where the line cuts the axis. Making the base perpendicular to the axis gives a circular right cone. The angle of the cone is determined by the angle between the line and the axis.

The central slices in each direction are triangles and all the others are hyperbolae (see figure 4-5 in chapter 4, "Properties and types of Surfaces"). You do not need to know anything about the geometry of a hyperbola, or even what one is, in order to construct each slice because its shape is obtained, point by point, by drawing and transferring lengths from the plan.

Drawing the base plan and the central slices

As with other models start by drawing a grid. Use a spacing of 1 cm and make the lines 10 cm long. This gives nine slices in each direction. As with other models, there is no slice on the end, since such a slice would not have any support.

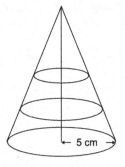 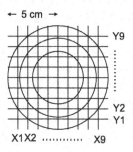

figure 5-10

The base of the cone is a circle, so draw a circle whose centre is the centre of the grid and radius 5 cm. Now draw two more circles whose radius causes them to fall between the grid lines. Use radii of 2.5 cm and 3.5 cm as shown in the right diagram of figure 5-10. The three circles correspond to points on the line which have been rotated about the axis. The outermost circle is the base of the cone and the other two form circles higher up where the circles would equal the radius of the cone at those positions, as shown in the diagram on the left. These circles are used to find the shape of a slice.

Now draw an isosceles triangle for each of slices X5 and Y5. These are the central slices, cross-section through the axis of the cone. The height of the triangle determines the height and angle of the cone. Use a height of 7 cm. The base of the triangle is the length of the lines of the grid (10 cm). Use the grid spacing to position the slots as described in chapter 1, "Introduction and basic techniques". This central slice is the left one in figure 5-13.

As well as drawing the triangles for slices X5 and Y5, you also need another version this central triangle to act as the cone cross-section for creating each of the other slices as shown in figure 5-11.

Drawing the shape of the other slices

The following diagrams and description refer to drawing the shapes of slice X3.

In figure 5-11, the triangular cross-section of the cone is RGF. Draw OF (the axis of rotation) perpendicular to the base. Then on the base line, starting from the mid-point, make marks which are equal to the radii of the circles you drew on the base plan grid. Distance OA is equal to the radius 2.5 cm, OC 3.5 cm and OG the base radius of 5 cm. Draw perpendiculars AB and CD up from these points to meet the sides of the triangle.

Extend the base of the triangle RG and draw a perpendicular at an arbitrary point J far enough from point G so that the new slice can be drawn around it. This perpendicular will act as the centre of the slice. To find the height of the slice, first mark a point H. The length OH is measured on the grid for the position of this slice. Since the grid is spaced at 1 cm intervals, this means OH is 2 cm in this case for slice X3. Draw HK perpendicular to RG and then draw line KM perpendicular to HK to intersect JM at M. You could also imagine that the triangle RGF is slice Y5 and line HK is slice X3 seen edge on.

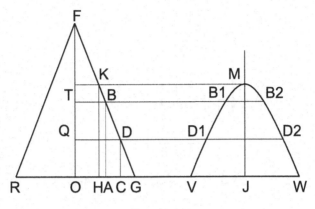

figure 5-11

Other points on slice X3 are found as follows, using the other two perpendiculars AB and CD, corresponding to the position of the circles drawn in figure 5-10.

1. First, draw the lines TB and QD parallel to the base of the triangle and find the positions of B1, B2, D1 and D2 as follows using figure 5-12.
2. Imagine that you are looking down onto the cone as in figure 5-12. The two bold lines show the length of B1B2 and D1D2 (remembering that X7 is the same as slice X3) because they are the intersection of the slice with the circle of revolution of points B and D. You can transfer these lengths to the drawing in figure 5-11 in one of two ways:
   - using the "paper ruler" method described in chapter 3, *Contours and profiles*, to mark the lengths on the edge of a piece of paper;
   - using a pair of compasses; place the point in the position where the central grid line Y5 intersects the grid line for the slice you are working with (a bold line in this case) and open the compass to where the circle cuts the line of the grid (the end of the bold line); then place the compass point at the point where, for example, TB intersects with JM and make two small arcs on TB at B1 and B2.
3. Find points V and W, the base length of the slice in the same way using the largest circle in figure 5-12.

4. Sketch the curve to complete the slice. In this case, it is possible to manage with just a few points to sketch the outline satisfactorily because the curvature is changing slowly, with almost a straight line between the points. If you feel you need more certainty, or are working with a more complicated profile, draw more circles on the base diagram.

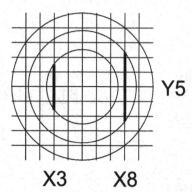

X3   X8   figure 5-12

Drawing the other slices and adding slots

Other slices are found in the same way. Because of the symmetry of the surface, there are only five different outlines to draw. The central slices for X5 and Y5 are the main triangle; X1, X2 and X4 (which all have symmetrical slices) are drawn in a similar way to X3. You may need some more circles in the centre to get smoother curvature at the top of the slice for X2 and X4. You will certainly need an extra circle close to the outside one for slice X1 since at the moment with the two circles drawn in figures 5-10 and 5-12 you are only able to find two points, the height and the width of the base. The innermost circles do not intersect this slice at all because they correspond to positions higher up the cone, above this slice.

The next stage is to add the slots. Mark lines vertically on the slice shape you have created. You know the grid spacing, so mark the centre of the base of each slice and mark of equal grid spacings from that centre. Figure 5-13 shows one set of slices. The other set is the same, but with the slots at the top.

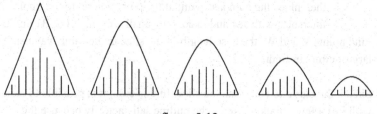

figure 5-13

Rotating about another axis

The position of the rotation axis makes a considerable difference to the shape of the surface of revolution. The surfaces in figure 5-14 were created by rotating an ellipse about an axis which is not the axis of symmetry (as was the case in figure 5-4).

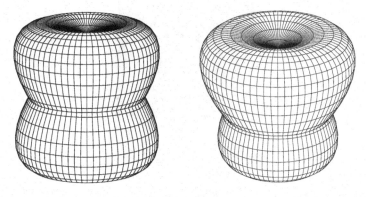

figure 5-14

The reason for the difference can be seen by looking at figure 5-15, which show different results for the cross section of a surface obtained by rotating all or part of an ellipse which has been tilted or translated by different angles before rotation.

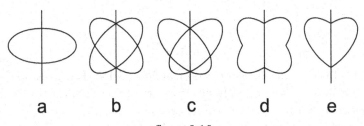

a   b   c   d   e

figure 5-15

Figure 5-15a shows the result without any tilting, when the ellipse is rotated about its minor axis, which results in the ellipse being superimposed on itself because of the symmetry. In figure 5-15b the ellipse has been rotated by 45° about an axis perpendicular to the plane of the paper before rotation. With figure 5-15c, the axis has been moved away from the centre, resulting in the waist of the surface being lower. The surfaces of revolution shown in figure 5-14 are the ones corresponding to figures 5-15b and 5-15c. Figure 5-15d shows the outside edge of figure 5-15b and is the only part of the result you would see if you made a Sliceform model to correspond to the left surface in figure 5-14. Mathematically, if you kept the complete ellipse, the surface for figure 5-15b and 5-15c would have an internal hole that is not visible in figure 5-14. Part of this could be realised by using the part of the curve of figure 5-15b shown in figure 5-15e to make another model.

fig 5-16

Figure 5-16 shows the Sliceform version of figure 5-15b. The slices are shown in figure 5-17. The top row shows them as they are created and because of the symmetry, only one set is required for each direction. The bottom row shows that they need special positioning of the slots and that the outside slice e has had to be cut into two parts, otherwise it would not be possible to slot the slice into the others. Slice e only fits into slices a to c since slice e does not intersect slice d. The slots for the two parts of slice e to fit into slices a to c are both from the outside. It is unusual in having slots in two directions on the same slice. Slice d has had the outside slots broken into two parts since the slice intersects either side of the waist. These have to be in the same direction because the slots have effectively had to be moved down.

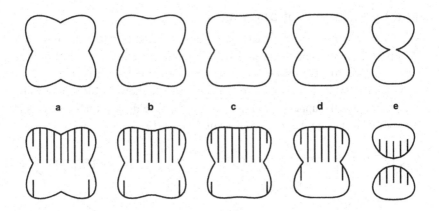

figure 5-17

Figure 5-18 shows the Sliceform model for figure 5-15e. Because the depression is on the axis, and there is no waist, the slices do not call for any special making of slots.

figure 5-18

More examples and suggestions

The methods described above for the cone apply equally well to other surfaces of revolution. There are many curves that can be used to produce surfaces of revolutions. The following are a few special examples.

The hyperboloid of one sheet

The hyperboloid of one sheet can be made as a surface of revolution. If you slice parallel to the axis the central slices are one piece but, depending on the hyperbola you rotate, at some stage the slices become two parts. Figure 5-19 show the central and one adjacent two-part slice and how the slots have to be adjusted so that the upper part of the non-central slice does not fall out. This is similar to the slots on the outside slices of figure 5-17.

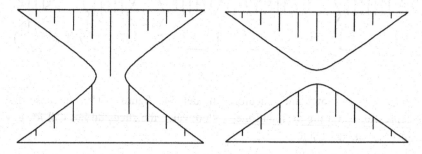

figure 5-19

and the complete model shown is figure 5-20 is not as robust as most Sliceforms, with the smallest slices still having a tendency to come out of their slots.

figure 5-20

An egg from an oval of Cassini

The family of Cassinian ovals in figure 7-14 in chapter 7 "Surfaces from equations - Algebraic surfaces", offers many possibilities for creating different surfaces using the different curves of the family. The innermost oval shape rotated about the horizontal axis gives rise to an egg as shown in figure 5-21.

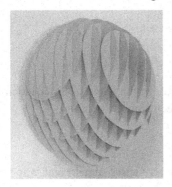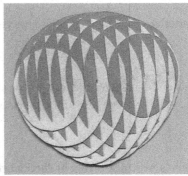

figure 5-21

The slices for this model are used as the example for making a different version of the egg by slicing along a direction perpendicular to the axis of rotation. This is shown in figure 9-15 in chapter 9, "Variations and Explorations".

A blood corpuscle from an oval of Cassini

The outermost curves from the Cassinian ovals family also yield interesting shapes. They have two axes of symmetry. The curves which are continuous when rotated about a vertical axis give rise to a surface which looks very much like a blood corpuscle as shown in figure 5-22.

figure 5-22

Lamé ovals and Supellipsoids

Lamé ovals are another family of curves, not strictly ovals, which yield interesting surfaces of revolutions. They are derived from an equation generalising the equation of the ellipse which has also given them the name superellipses.

The cartesian equation of an ellipse has the form:

$$\frac{x^2}{a^2} + \frac{y^2}{b^2} = 1$$

where a and b are the major and minor axes of the ellipse. This equation can be seen to be a special case of the general equation:

$$\frac{x^n}{a^n} + \frac{y^n}{b^n} = 1$$

The curves were discovered by the French physicist Gabriel Lamé who wrote about them in 1818. Varying the exponent n gives a range of interesting curves, some of which are known in other guises (like the astroid), but are usually known as Lamé ovals if considered in this way. If a is 1.618 and b is 1, some curves with different values of n are shown in figure 5-23.

figure 5-23

If the values of n are greater than 1 then the curve is convex. A value of 1 gives a rhombus (with straight lines) and less than 1 concave, astroid like, curves result.

The Danish writer, recreational mathematician and inventor Piet Hein used the curves having a value of n greater than 2 (like the bolder lined one in figure 5-23) as a source of design for roads and the idea was also taken up in furniture design. He called the curves super-ellipses because they look like ellipses but are more rectangular. He thought the superellipse with a value of n as 2.5 was the most beautiful one and created an interesting surface of revolution which he called a superegg. Wooden supereggs were a popular "executive" toy in the 1970s. They balance on their end easily in whatever position they are placed. For the full story, see Martin Gardner's "Mathematical Carnival". The Sliceform version is shown in figure 5-24 with the central slice shown in figure 5-26a.

The superegg as a Sliceform model is not quite as easy to balance as a solid version because it is not a solid object. The balancing properties of the superegg variations made by deforming it using a Sliceform version would make an interesting research project.

 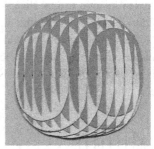

figure 5-24

Such curves offer many possibilities for creating surfaces, one of which is to take part of the mathematical curve and use that as the starting point. For example, taking a quarter of the curve in bold in figure 5-23 and rotating it about an axis which joins its ends (which is the same as saying the line forming the rhombus) yields a totally different surface. This is shown in figure 5-25 and the central slice is shown in figure 5-26b.

figure 5-25

This suggest another possibility. The slice can be cut in half and the two halves assembled to give a new slice. This is illustrated in figure 5-26. The first slice (figure 5-26a) is the central slice for the model in figure 5-24. This has two axes of symmetry and so would give the same shape if this idea were applied. However, figure 5-26b, the slice for the model in figure 5-25, only has one axis of symmetry. Cut and reassembled it gives the central slice in figure 5-26c.

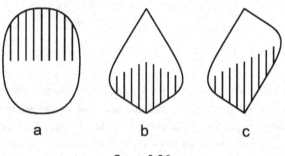

a   b   c

figure 5-26

Applying this method to all slices gives the Sliceform shown in figure 5-27.

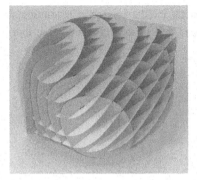

figure 5-27

More models from the same surface design

The possibilities for Sliceform surfaces of revolution is endless. In addition to the method used for the model in figure 5-25, there are two ways that further models can be made. One is to take one set of slices from the surface of revolution and make a different set of slots, perpendicular to the axis. Then create a second set of slices which are circles having radii that match the size of the slots. The circles correspond to the revolution of the edge of the curve. This is described, with an example using the Cassinian egg surface in chapter 9, "Variations and Explorations". The effect is shown in figure 9-15, with figure 9-14 as the original egg.

Another way to create more models from slices of surfaces of revolution is also described in chapter 9 in the section "Three sets of slices from two". Many models are symmetrical in that the two sets of slices defined by the construction are the same shape, but in general there are three different sets of mutually perpendicular slices possible. Taking them two at a time gives three models. Applying this to the slices for figure 5-26c does *not* give the same model as figure 5-27.

Different approaches

Chapter 7, *Surfaces from equations - Algebraic surfaces*, has two different descriptions for creating surfaces of revolution using algebra. The first uses the catenoid surface as an example and is in the section describing parametric equations. The second method, in the section "Sculpting with equations" has a different approach which does not strictly yield a surface of revolution since a curve is not rotated parallel to the axis, but it does give a surface that has rotational symmetry.

Chapter 6
Ruled Surfaces

This chapter describes methods for creating Sliceform models using ruled surfaces.

Generation of surfaces from straight lines

Monge's description of a curved surface in chapter 4, "Properties and types of Surfaces", includes surfaces generated by the movement of lines. Where the line in motion is a straight line, then the surface is said to be a ruled surface. The set of lines are called generators. Unless the line is moved in a plane, the surface is curved.

To create a Sliceform model, using the generators to define edges of slices, they must relate to the grid. In many cases, the lines intersect one another so that defined slices are not possible. This does not mean that examples of the surfaces cannot be produced as Sliceform models, simply that they have to be sliced in other ways. Some of the descriptions below are to show the way some surfaces can be defined as ruled surfaces, to explain these situations.

In order to make a Sliceform model, there needs to be solid substance to slice. The restriction on making the generators as edges of slices means they just form a surface and not enclose a volume, so it is usually necessary to create a slice by some other method with the line of the ruled surface forming one edge of the slice. The easiest way to do this is to use the ruling line with the base line of the grid to form a plane which constitutes the slice.

Extrusions of curves as described in chapter 3 "Contours and profiles", are also examples of ruled surfaces. They can be thought of as a curve in a plane with one end of a line perpendicular to the plane which then moves along the curve.

The simplest ruled surface - a cone

A cone is produced when a line is moved so that one point of the line remains fixed and another point of the line traces a curve. If the curve is a circle and the fixed point is on an axis perpendicular to the plane of the circle and through its centre then a right circular cone is created. Note how the cone extends both sides of the fixed point.

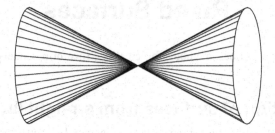

figure 6--1

A cone where the base curve is a conic (an ellipse, parabola or hyperbola) is a cone of the second degree. Other curves can be used as the base curve to create more sophisticated surfaces. None of these is suitable for creating as a Sliceform model using the ruled surface generation directly, since all the slices would go though the central point. A model of the cone is described in chapter 5 "Surfaces of revolution", where it is sliced using planes parallel to the axis.

Cylinders and hyperboloids

If a cone has the fixed point moved to infinity, then the ruled surface becomes a cylinder. So, if a line segment is rotated about an axis, with the line parallel to an axis and away from it, then a cylindrical band is generated. Each point of the segment forms a circle in space. If the line is not parallel to the axis, each point of the segment still forms a circle, but points corresponding to each line on these circles are no longer parallel to the axis of rotation; the surface then becomes a hyperboloid of one sheet. You can think of this as a cylinder with a "waist" or a cone where the central point has opened out into a circle.

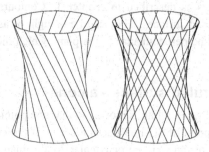

figure 6--2

There are two sets of generators in the surface, one a reflection of the other as shown in the diagram on the right of figure 6--2.

Like the cone generated as a ruled surface, this method is not suitable for creating a Sliceform model. A model of a hyperboloid of one sheet is best constructed as a surface of revolution (see chapter 5 "Surfaces of revolution"), using one half of the hyperbola as the curve being rotated and is shown in figures 5--19 and 5--20 in chapter 5.

The helicoid as a ruled surface

A helicoid is like a screw and is a three dimensional version of a spiral. Imagine a line segment one end of which slides up an axis while the other end rotates about the axis. If the rates of sliding and rotating are constant, then spacing of the helicoid is constant and the result is shown in figure 6-3.

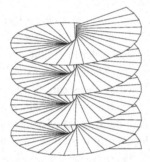

figure 6--3

The helicoid generated in this way does not obviously suggest ways to create a Sliceform model. The "open" nature of the surface where no volume is enclosed makes it difficult to find substance for the slices.

The Möbius band as a ruled surface

If a line segment is rotated about an axis, but is rotated parallel to the axis and away from it, then a band is generated. Each point of the segment forms a circle in space. If such a band is cut and the ends stuck together after giving them a half twist then this gives famous Möbius band which can also be generated as a ruled surface.

Consider the circle which formed by the central point of the segment in generating the band. If the line segment is rotated as it goes around the circle so that after one circuit of the circle it has turned through 180°, then this will generate the half twist. The result is the Möbius band as a ruled surface.

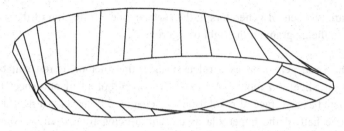

figure 6--4

As with the conventional helicoid, the twisting causes problems with trying to make it into a solid form that can be sliced.

What types of ruled surface are suitable for Sliceforms?

To create a Sliceform of a ruled surface, we need to have a surface that is non-intersecting, otherwise it needs to be constructed as a totally different type of surface using string for the ruled lines in order to see the internal structure. As a general rule, the easiest way to design Sliceforms is to slice a cube or cuboid. Conveniently, a large number of ruled surfaces can also be made by joining points on a cube or cuboid. Such ruled surfaces when projected onto a particular plane (the base of the cube) define a set of parallel lines. That is not to say that a ruled surface which does not fit this criterion cannot be made as a Sliceform, just that another way to slice must be used which does not rely on the ruled lines, as in the hyperboloid of one sheet as described above.

The Hyperbolic Paraboloid – a ruled surface

The easiest ruled surface for creating a Sliceform model is the hyperbolic paraboloid. The following description shows how the ruled surface can be the top of a solid object, and how variations can also be created to approximate just the actual surface. This surface is important architecturally since it has been used to create a curved surface by using a set of flat rectangles. For example, the Commonwealth Institute in Kensington in London has such a roof made out of sheets of copper.

The lines in the ruled surface

If you draw the diagonals of the faces of the cube, some of them meet at the corners and some are parallel to one another. There are also a number which never meet in space and are not parallel. Such lines are called skew lines. The ruling lines of the hyperbolic paraboloid are drawn by joining points on a pair of skew lines such as AF and CH in figure 6--5.

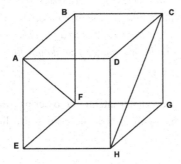

figure 6--5

This gives a spatial quadrilateral called a skew quadrilateral. Opposite sides of this quadrilateral are not parallel and if extended would never meet, as you can see from figure 6-6.

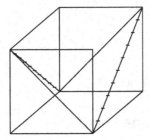

figure 6--6

First divide each of these lines into an equal number of parts, say ten. Then join the ends of the lines by another pair of skew diagonals. Now join the points on each of the marked diagonals, to build up the surface of the hyperbolic paraboloid as shown in figure 6--7 which is two different views of the surface acting as a stereoscopic pair.

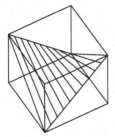
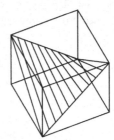

figure 6--7

[Note: Pairs of images like this are stereoscopic pairs which enable you to see the surface in three dimensions. Place the page 8-9 inches from your eyes; then cross your eyes slightly so that you see three images. Focus your attention on the centre one which should appear as a three dimensional image. This may take a few seconds, and once you can do it with one image, you should find it easier with others.]

Because of the symmetry of the skew quadrilateral, the other pair of sides can be joined in the same way. Figure 6--8 shows how these lines also lie in the same surface.

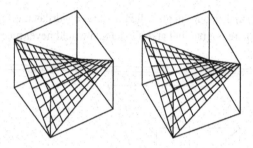

figure 6--8

Where does the name come from?

If you slice the surface with a plane in the two perpendicular directions through the corners of the skew quadrilateral, then the curve on the slicing surface is a parabola. If you slice across the surface using a plane which is perpendicular to these two parabolas, then you get a hyperbola. There is also another way to see the parabola.

Suppose the cube used to generate the surface is rotated. As it becomes more upright, then the lines cross one another and you can start to see the parabola appear as a cross section.

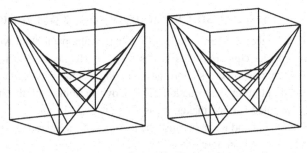

figure 6--9

When you look (with one eye) with a vertical edge of the cube near to you, then the parabola appears symmetrical and in two dimensions appears as a envelope of a parabola; so that the three dimensional view of figure 6--9 becomes the form at the left of figure 6--10 which is like the diagram commonly used to draw the parabola in curve stitching. The diagram at the right of figure 6-10 shows the points of the cube labelled as in figure 6--5. Because you are looking at the cube with a vertical edge towards you, the lines BF and DH appear on top of one another.

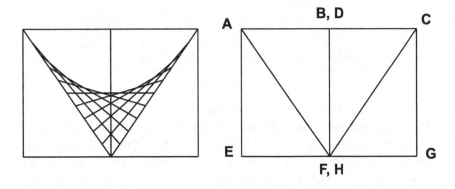

figure 6--10

You see the lines crossing even though only a set of lines touching one pair of skew lines has been drawn. If both sets of lines are drawn, when you look from along an edge of the cube, the lines from the two sets come on top of one another. This symmetry makes it easy to derive a construction for the slices of the Sliceform model and simplifies the design.

A simple construction for the hyperbolic paraboloid

One way to make the Sliceform model of the hyperbolic paraboloid would be to generate it using its equation. This is described in chapter 7 "Surfaces from equations - Algebraic surfaces" and the following technique is just a geometric interpretation of plotting the equation. The above definition of the hyperbolic paraboloid as a ruled surface makes it easy to construct the slices. It is quite an easy model to cut out and make because the top lines of the slices (the ones which form the surface) are all straight lines.

To determine the shape of the slices, draw squares corresponding to two opposite sides of the cube and mark the diagonals with divisions corresponding to the ends of the lines to be joined. Draw lines from these diagonal divisions down to the base of the square. These lines form the ends of the slices.

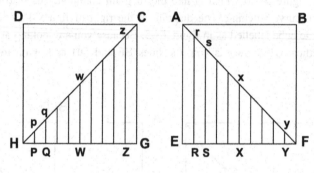

figure 6--11

In figure 6--11, letters A to H correspond to the letters in figure 6--5. The ends of two slices have been labelled Pp and Qq, Rr and Ss. The shapes of these slices shown in figure 6--12 are drawn as follows. PR and QS are the length of the edge of the cube. Opposite edge lengths Pp and Rr (and similarly Qq and Ss) are obtained from the diagrams in figure 6--11. Joining pr and qs completes the slice shape.

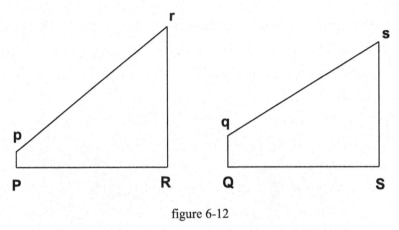

figure 6-12

When you drawn the outlines for each of the slices using the other lines in figure 6--11, the next stage is to construct the slots. Mark the positions using the same spacing as was used in the diagrams in figure 6--11 (as shown on slice pPRr at the left of figure 6--13) and measure half way to find the length of the slots (as shown on slice qQSs on the right in figure 6--13).

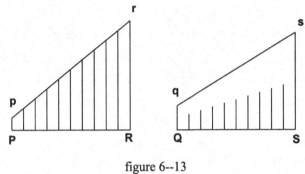

figure 6--13

There is no need to measure individual lines in this case; simply measure the mid-points of the end lines (pP and rR for example) join these points and where the resulting line cuts the vertical lines is the mid-point of each one.

There is also no need to produce a complete drawing of slices from all lines since there is a symmetry about the mid-lines (xX and wW in figure 6--11) forming the middle slice . This slice is a rectangle if the line is the mid-line of the square . The slice formed from pPRr is the same shape as the one from zZYy turned over.

This symmetry of the surface means that the slices for the other direction are exactly the same shape. However, the slots are different. Instead of being cut from the base up as on the slice qQSs, they must be cut from the line at the top down to half way.

The model when constructed looks like this:

 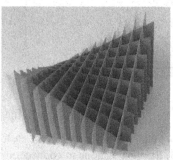

figures 6—14a and 14b

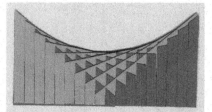 

figures 6-14c and14d

As you can see in figures 6-14c and figure 6-14d, the two directions of flattening yield totally different results.

Closer to the surface

The hyperbolic paraboloid has been modelled above as a surface which can be thought of as cutting away part of a cube. Geometrically it is just a surface without anything either side. The following is a method to get closer to making a model which just shows this aspect of a surface.

Consider a parallel surface below the one defined by joining points on diagonals of the cube, so that the slices become parallelograms. They are constructed from the slices as shown in the figure 6-15 which is a modification of figure 6--12. The height of each slice is the length pP. The same height is used in the other slices, for example on slice qQSs on the right.

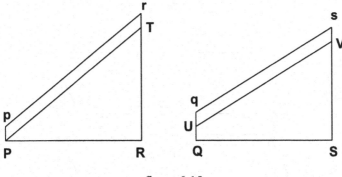

figure 6-15

Making a model using these slices does not work very well if you try to use the normal way of constructing the slots: half way up in one direction of the grid and half way down in the other. The model tends to fall apart. This can be overcome by alternating the slots like those in figure 6--16:

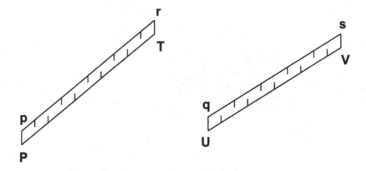

figure 6-16

When you construct the model, you do so by weaving them since the slots go alternately over and under in both directions. This weaving holds the slices together like a woven textile or basket.

A simple and very pleasing Sliceform of this type can be made with just six slices (three in each direction). It is very quick to make. There are only two slices to design because of the symmetry. A complete set for one direction is shown in figure 6-17.

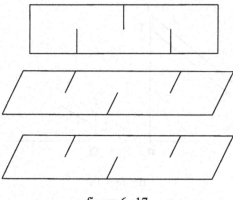

figure 6--17

When building the model, slot the two rectangles together and then add two of the other pieces (one from each direction) together like this:

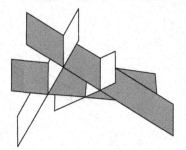

figure 6-18

Ensure that the symmetry at the top corner is correct to give the butterfly wings effect. Because the model can be put together in different ways and it does not fold flat very well if assembled incorrectly, I often use it as puzzle.

This model is one you really do have to make to see its full beauty. The photographs in figure 6-19 showing the flattened and open form do not do it justice.

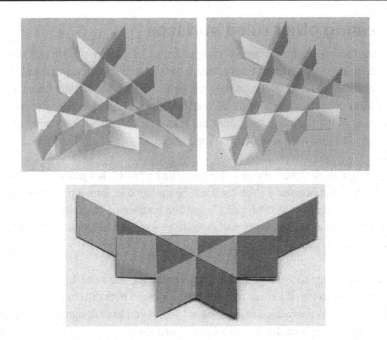

figure 6-19

Because this Sliceform is made up of only six pieces, and the edges are straight lines, it is easy to make as a larger sculpture. Figure 6-20 shows the work of Richard Ahrens. Such a sculpture will not fold flat because of the thickness of material (corrugated board in this case), so the slots all need to be in the same direction and not alternating as for the paper model.

figure 6-20

Creating other ruled surfaces

Taking the hyperbolic paraboloid a stage further, a number of different models can be made using different lines or curves on the ends of a cube or cuboid. They can be different on each end, or they can be the same with different orientations. There are many possibilities to be creative, since very few ruled surfaces are recorded in the literature apart from the ones described at the beginning of this chapter. The following are just a few suggestions.

In these examples, the cube or cuboid is called a "box", not just to simplify the description, but because a cube can itself be modified to form a parallelopiped as described in chapter 8, "Polyhedra", and the number of possibilities multiplied since there are more symmetry variations to be explored. Examples are shown at the end of this chapter.

When designing such models, the slices containing the ruled lines are easy to construct. However, the slices may not be the same in both directions, so you may have to resort to the method in the section "Where you have designed slices in just one direction" in chapter 1, *Introduction and basic techniques* to define the shape of the second set of slices. (The cycloid model in chapter 3 also describes the practical application of this technique.) However, look at the way the curve transforms from one end of the box to the other. It may do so linearly, or you may be able to spot a symmetry like the way the hyperbolic paraboloid has the same slices in two directions.

Conoids

A ruled surface that is produced from a curved and a straight line is called a conoid. The line is often called the directrix. A variety of Sliceforms can be created using the techniques described above for the hyperbolic paraboloid. The simplest is to replace one of the skew lines by a quarter of a circle.

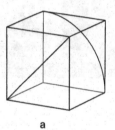 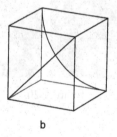 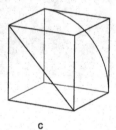

a   b   c

figure 6--20

Because of the non-linearity of the circle, there are two possibilities for replacement as in the two diagrams at the left of figure 6--20. The diagram at the right of figure 6--20 is the equivalent of the two diagonals being parallel and is a different variation. Two parallel lines would generate a simple triangular prism, but the convexity of the arc of the circle makes for a curved surface rather than a plane. This suggests a different model (a conoid "wedge") derived from a line and a complete circle, which is described below. The left and right diagrams of figure 6--20 form part of this conoid wedge.

The design of these models requires one set of slices to be drawn then you need to use the method as described in the section "Where you have designed slots in just one direction" in chapter 1, "Introduction and basic techniques" to define the shape of the second set of slices .

The model corresponding to figure 6--20a is shown in figure 6—21 and to figure 6--20b in figure 6--22.

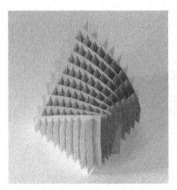 

figure 6—21   figure 6--22

The conoid wedge

The diagram at the right of figure 6--20 can be seen as part of a design for a larger model. Take a cube with a circle at one end and then use the mid-line of the opposite side as the directrix as shown in figure 6--22. This gives a wedge shaped surface. One set of slices contains the ruled lines, but in the other the circle gradually gets transformed into a line through a succession of ellipses. This makes it easier to design this set of slices.

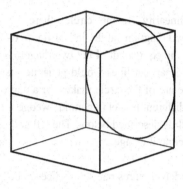

figure 6--23

The photograph in figure 6--24 shows the model resting on the circular face. It suggests an architectural design. It also shows interesting patterns when folded flat.

figure 6-24

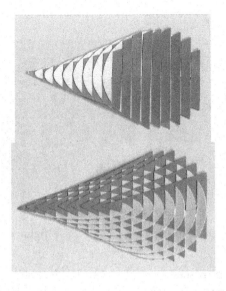

Curve to curve variations

The conoids in figure 6—20 suggest that variations can be made just using arcs of circles, that is with no straight lines.

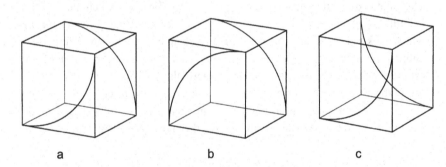

figure 6-25

In terms of a surface, the centre and right diagrams in figure 6-23 are the same. Each is a rotation, by 180°, of the other. However, as Sliceform models, they are different because the slices are not the same. The two added together form a cube. However, the model for the right hand diagram has such small slices that it is very difficult to make.

The model corresponding to the left diagram of figure 6-25 is shown in figure 6-26 and to the central diagram in figure 6-27.

 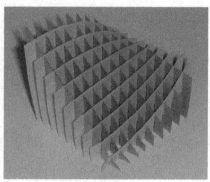

figure 6-26   figure 6-27

Note how the circle "pushes" the surface up while they still have a saddle shape. The surface in figure 6-27 is, however, reversed relative to the other models in this series of variations.

Using a different "box"

The cube is only one starting point. It is a regular "box" on which the structure is built. Using a cuboid, that is a cube that has been stretched, provides models which look very similar until you start to examine them closely. The lack of cubic symmetry means that folding the model flat results in a completely different pattern. The design process is exactly the same as for the cube. In the following photographs of the models, note that the variations in the patterns when the models are flattened is much greater than the cube based model shown in figure 6-14.This is because of the different symmetry.

Figure 6-28 shows a rectangular box instead of a cubic one which gives different patterns for the two ways of folding flat. Compare the results with figures 6-14c and14d

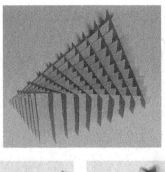

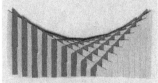 

figure 6-28

Figure 6-29 (with the model shown in figure 6-30) shows a cubic box, but with the lines between which the rules are drawn varied, so the line on the back wall is parallel to the base.

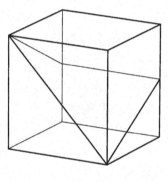 

figure 6-29   figure 6-30

Chapter 8, *Polyhedra* describes making a parallelopiped as an alternative to the cube. These are interesting because it is sometimes possible to make different models from the same set of slices which not only look different as open models, but also behave in widely different ways when folded flat. The first example is shown in figure 6-31.

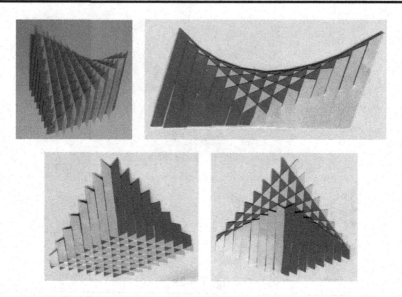

figure 6-31

The second example (figure 6-32 below) looks to be a very similar model, but the flattened forms are quite different.

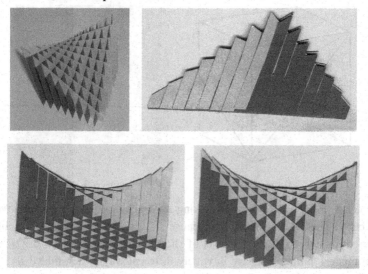

Chapter 7
Surfaces from equations - Algebraic surfaces

This chapter describes methods for creating Sliceform models which are defined algebraically. This is possibly the most mathematical chapter. If you find the creation of the Sliceforms using algebra rather than geometry not as interesting as other chapters, you may be tempted to skip the chapter, but at least take a few minutes to wonder at the wide array of surfaces which have been discovered by mathematicians.

Introduction

Most of the Sliceforms described in this book are defined by physically performing some geometrical action in space. This is historically how geometry became a branch of mathematics, since from its name, it was originally a science of measuring the earth. Such "pure" geometrical methods were supplemented with the recognition that that algebra can also be used through the use of systems of coordinates and functions. In looking at geometry from this algebraic or analytical point of view, some models are easier to make since their definition by purely geometrical actions may not be obvious and have only been studied by their definition as algebraic surfaces.

The heyday of the study of algebraic surfaces was at the end of the nineteenth century up to the first world war. They were studied partly as a way of visualising the mathematics of the functions they represent. Many museums and mathematics departments of older universities have collections of plaster and wooden models (often stored away) built for teaching or research purposes. Some of these were built up as sets of slices of wood because this was a practical way to make the model. Plaster models were often made by pouring the plaster into a Sliceform model made from cardboard. Gerd Fischer's book on mathematical models is a record of many of the ones in museums.

When mathematicians started modelling surfaces with computers, they used methods that were similar to the ones that had been used for plotting curves on paper by calculating a series of curves which moved over the surface and produced a mesh. The mesh looks like a woven net which has been draped over the surface. This method was also used by computer artists before they were able to paint the surfaces (see Prueitt and Reichardt). In computer graphics, even now, a surface is built up of an underlying mesh which is then painted or has a texture added.

With mathematical surfaces, the mesh, as can be seen in the illustrations later in this chapter, is built up from curves which lie on the surface being represented and there are two sets of curves at right angles to one another (corresponding to the x and y directions of the cartesian coordinate system). It is only a short step from these mesh curves to produce slices which can then be used to make a Sliceform model. There are also other coordinate systems and while these may define a surface elegantly, they may not be easy to plot in the cartesian system.

The mesh drawn for a surface called the monkey saddle surface (which is a model described later in this chapter as a Sliceform) looks like this:

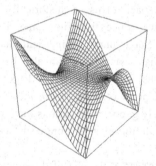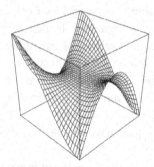

figure 7-1

[*Note: Pairs of images like this are stereoscopic pairs which enable you to see the surface in three dimensions. Place the page 8-9 inches from your eyes; then cross your eyes slightly so that you see three images. Focus your attention on the centre one which should appear as a three dimensional image. This may take a few seconds, and once you can do it with one image, you should find it easier with others.*]

The monkey saddle gets its name because it is easy to imagine a monkey sitting facing to the right with its legs going either side as the surface comes down, and its tail flowing down the valley at the back.

The surface defined has no solidity. To create a Sliceform, the curves of the mesh are cut along the top of the slices. The slices then act as the "support" for the surface curves in the way that a piece of paper acts as a support for a drawn curve. This leaves open the question as to what to do about the cut at the bottom of the slice. Consequently, there is scope for creating different models from the same surface by the way the other edges of the slices are drawn. Placing the surface in a box means the other sides are the slice are straight lines. So a typical slice for a surface, with the slots cut in it, looks like this:

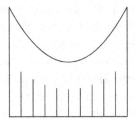 figure 7-2

If you look at the mesh diagram for the surface, note how there are a large number of curves on the mesh to enable the surface to appear smooth. When you make Sliceform models, the curves on the slices have to be smooth, but you do not need many slices to make a smooth surface. Your eye and brain construct the surface very easily.

Representing surfaces as equations

We are used to representing curves in the plane by different types of equations, for example by using cartesian or polar coordinates. We can also use similar coordinate systems to represent surfaces, but we need three coordinates, rather than the two in the plane, since we are working in space. Not all coordinate systems are suitable for creating Sliceforms. The easiest to use are three dimensional cartesian coordinates. In a plane the x and y coordinates define a position relative to an origin. It is convenient to keep these for the base grid of a Sliceform and to use the z coordinate for the height of a slice.

In determining the height of the slices, you need to make sure that you work from the slice that has the minimum for the surface. Otherwise it is very easy to miss part of the surface. Also, you need to make sure that the base of the slice is a sufficient distance from the curve. Slices like the one in figure 7-3 will easily fall into two pieces when the slots are cut.

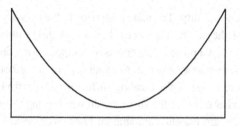

figure 7-3

Use the slice which has the minimum for the curve as a standard to know how deep to make the base. Make sure that you have sufficient depth to create the slots and provide sufficient strength for all slices. The curve in figure 7-3 needs to be as far away from the base as in figure 7-2. Remember that half of the height will be taken up by the slot; making this height too small will weaken the slice. This may not too be important for the appearance of the final model, but short slots will cause problems during assembly and flexing.

Another consideration in making Sliceform models using equations is that that slicing is always easiest along directions parallel to the axes. This is because calculations are almost always achieved by setting one of x, y or z to a constant for the position of the slice.

Explicit equations

Some equations of surfaces are easier to plot than others. For example, an equation like:

$$z = x^2 + \frac{3}{y^2}$$

is easy to work with because if you know a corresponding pair of x and y values you can calculate the z value directly. This is known as an explicit equation. To form a slice, the procedure is to define the direction of a slice as constant (for example, choose a value of y) and vary the other coordinate (moving in the x direction moves along the slice) and so produce a curve showing the position of z with respect to x along the y slice direction. Taking a discrete set of y values gives one set of slices. Repeating this with a discrete set of x values and varying y gives the other set of slices.

The discrete values of x or y also define the positions for the slots in each set of slices. The slots are half the height of the slice at each point as with other Sliceforms.

Example of creating a sphere

The following steps show how this could be used to create the slices of sphere by using an equation rather than the geometric method in chapter 2.

1. Using Cartesian system of coordinates, a sphere, with centre at the origin, has the equation:

$$x^2 + y^2 + z^2 = r^2$$

where r is its radius.

2. Decide on two directions which define the two sets of slices, for example the xz and yz directions and also decide the radius of the sphere.

3. For slices parallel to the xz plane choose y values corresponding to the position of the slices. For example if the radius of the sphere is 10 units, let y take the values -8 to 8 in steps of 2 which gives a total of 9 slices in this direction. Because of the symmetry, you only have to use values from 0 to 8 to obtain all the different shapes of circles.

4. Set y to each of these values in turn in the above equation, to reduce it to a planar equation which is a circle whose radius is $\sqrt{(10^2 - y^2)}$ if the radius is 10 units.

5. Plot a set of circles with these radii and mark the slots to correspond to the positions in the x direction, with x taking values -8 to 8 in steps of 2. Work from the centre of the circle in placing the slots.

6. In other models described in this chapter, you would usually have to repeat the process for slices parallel to the yz plane, but in this case the symmetry of the sphere means you do not need to do so.

This is a simple equation, and is given at this stage to show the general principles. A more complex example is given later in the chapter to draw a Monkey Saddle surface which is both not a closed surface and also has varying curvature. Both of these mean more decisions have to be made in designing the slices. Further examples of explicit equations and computer images are shown later in the chapter, with ideas on how to create the models.

Implicit equations

Equations where x, y and z are not related in such a simple way, such as:

$$(x + y^2)^2 + (z - y^2)^2 = a(k - y^2)^3$$

(where a and k are constants) are much harder to deal with. They are known as implicit equations. Solving implicit equations is an art. They may have an easy solution in special cases. One technique in solving them is to make x, y or z constant and see if the equation is solvable to give an explicit equation. You may also be able to solve such an equation using a computer program, or to plot the series of equations for each slice using special plotting software. This is described in chapter 10 "Using the Computer". Once you have a set of equations plotted for each slice the construction of the Sliceform model is the same as for explicit equations.

Examples of surfaces with implicit equations

The following examples show some surfaces that are defined using explicit equations. Not all are suitable for making Sliceform models, usually because they are often more complex. Trying to find suitable equations can be quite time consuming. Special techniques for constructing the slices as described in chapter 10, "Using the Computer", and these are more complex than just plotting slices of an explicit equation. So it is advisable to use software for visualising the surface and deciding how to slice. This also helps in varying coefficients in the equations to help to make a surface suitable for modelling. As with other methods of plotting equations, the range selection defines part of the surface, and this may be suitable for making a model, whereas the whole surface may not, especially if it splits into separate pieces. Ranges for each equation are given with the description of each surface.

In the following diagrams of the surfaces, each one is shown as a stereoscopic pair, like figure 7-1. See the description under figure 7-1 for more details on viewing stereoscopic pairs. Not all surfaces shown as computer graphics have been made into models, indeed some are not suitable.

Double tear drop

This surface (figure 7-4) is one of those surfaces that is a mixture of convexity and concavity, and so finding the position to cut the slices so that the peaks are seen makes it very difficult to make a good model. The extra work involved in creating slices from implicit equations has certainly deterred me from attempting it, although it looks very aesthetic. It is a surface I discovered/invented have given it a name which says it looks like a pair of tear drops coming together.

figure 7-4

Its equation is:

$$(x + y^2)^2 + (z - y^2)^2 - 0{\cdot}1(3 - y^2)^3 = 0$$

The plotting range for x, y, and z is from -3 to +3.

Tear drop

This surface (figure 7-5) looks like a drop that is breaking away from a surface.

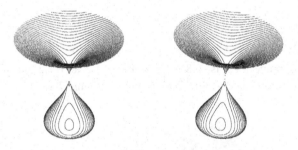

figure 7-5

Its equation is:

$$0{\cdot}4y^5 + 0{\cdot}7y^4 - (x^2 + z^2) = 0$$

The plotting range for x and y is -2 to +2, and for z -1.8 to +1.5.

The drop appears to have a peak and the whole surface has rotational symmetry, so this would not be a problem in making a model. In fact the two parts are joined by a line, and the model needs to be made as two separate objects, the drop shape and the funnel shape.

Heart

This is an example of an equation which defines a surface that would be difficult to sculpt using other techniques. There is a curve called a cardioid (which means heart shaped) but this is rounded at the point opposite the cusp. A Sliceform made using the cardioid to generate a surface of revolution looks more like an apple and was used as such in my book of cut-out models.

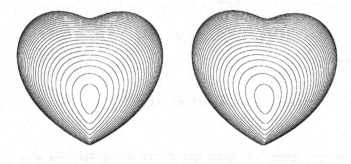

figure 7-6

Its equation is:
$$(2x^2 + y^2 + z^2 - 1)^3 - (0 \cdot 1x^2 + y^2)z^3 = 0$$
The plotting range for x is -0.8 to +0.8, for y -1.2 to + 1.2 and for z -1.3 to +1.3. The Sliceform model of the heart is shown in figure 7-7.

figure 7-7

Kazoola surface

I do not know why this surface is so named, but is has many variations depending on the values of the coefficients. Not all are suitable for making models. Moreover, normally a Sliceform model only shows an external surface. There may be an internal hole, or a splitting of the surface into two parts where one part is nested within the other. Its equation is:

$$d + cx^2y^2z^2 - a(x^2 + y^2 + z^2) - b(x^4 + y^4 + z^4) = 0$$

The case shown is with the coefficients, a = 1.875, b = -0.4, c = 1, d = 1.62 as shown in figure 7-8,. Figure 7-9 show the effect of slicing away part of the surface so that the central part is visible through the hole. The plotting range for x, y, and z is from -2.5 to +2.5.

The external surface is like a cube which has indents in the faces and at the vertices, giving the overall appearance of the polyhedron known as a cuboctahedron.

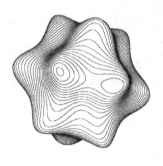 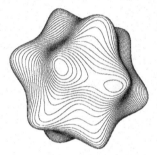

figure 7-8

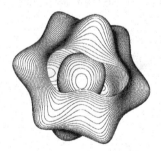 

figure 7-9

In creating the slices for the model (figure 7-10), the slices of the internal part are ignored.

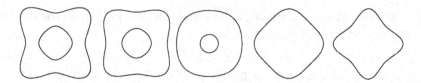

figure 7-10

The cuboctahedral shape is evident in the model (figure 7-11) and the pattern of the flattened shape is interesting too.

 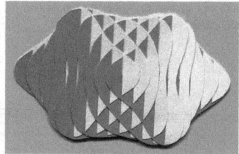

figure 7-11

Elliptical Vase

It is possible to take a surface of revolution and squash it in the direction perpendicular to the rotation axis and so get surfaces which are elliptical in cross section instead of circular. This elliptical vase surface (figure 12) , however, is generated from an implicit equation.

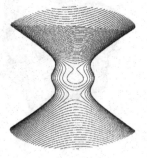 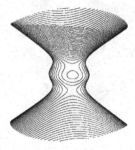

figure 7-12

Its equation is:

$$z^2 - \ln(z^2 + 0{\cdot}4) = 0{\cdot}5x^2 + y^2$$

The plotting range for x is -4 to 4 and for y and z is from -3 to +3.

Three focus surface

In chapter 10, the section "Programming techniques for curves and surfaces defined logically" describes how to plot slices for surfaces like those described in chapter 3 (Figures 3-14 and 3-15) which are defined logically (analogous to the definition of the ellipse as the curve which is the locus of points where the sum of the distances to the foci is constant). In many cases such surfaces can be expressed as implicit equations. This equation defines the surface shown in figure 7-13. It is defined as follows. Three foci are defined in space on the plane through the x and y axes with coordinates (a, b, 0), (c, d, 0) and (e, f, 0). The surface is defined as the locus of points where the product of the distances of each point on the surface to these three foci is a constant. The distance of a point (x, y, z) from the point with coordinates (x_1, y_1, z_1) is:

$$\sqrt{(x - x_1)^2 + (y - y_1)^2 + (z - z_1)^2}$$

so the equation of the three focus surface is:

$$k = \sqrt{(x - a)^2 + (y - b)^2 + z^2} + \sqrt{(x - c)^2 + (y - d)^2 + z^2)} + \sqrt{(x - e)^2 + (y - f)^2 + z^2}$$

where k is the constant for the product. In the example shown in figure 7-13, the three points have coordinates (-0.5,0), (0.5,0,0) and (0, 1,0) and the constant (k) for the product is 0.3. Varying the constant k causes the three lobes to become more rounded or for the surface to split into a number of pieces.

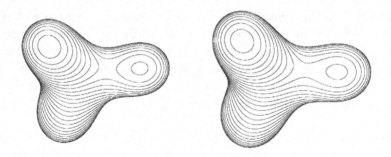

figure 7-13

A similar set of surfaces could be defined for where the sums of the distances to the points are constant. This surface definition is analogous to the definition of an ellipse as the locus of points where the sum of the distances from two points in a plane is constant. Varying the constant gives rise to a wide range of surfaces. Moreover, if the there are only two points and the locus is defined as a surface, then the surface is an ellipsoid. Figure 7-14 shows the surface formed where the three focus point are the vertices of an equilateral triangle.

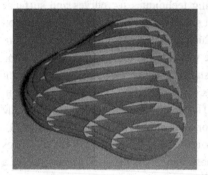 

figure 7-14

Parametric equations and polar coordinates

These are sets of equations in which the x, y and z coordinates are defined by a pair of parameters, allowing any point to be specified, without a specific equation connecting x, y and z.

Parametric equations are most often found where surfaces are described in spherical polar coordinates. These parametric equations are not suitable for creating Sliceforms unless a parameter can be fixed so as to define a plane of the mesh. Because polar coordinates define points from a centre (the pole), then they usually lead to a set of planes around an axis, which makes them impossible to use in creating a set of slices on a mesh for slicing along the lines of a grid. Another method of modelling such surfaces is given at the end of chapter 9 "Explorations and variations".

This is easily illustrated in the case of a sphere which can be represented parametrically by equations which define the points on the sphere in terms of two angles. Parametric equations can be used to create a mesh, but if the parameters involve polar coordinates, then the shape of the mesh reflects this. A mesh for a sphere defined by angles θ and ϕ (the parameters which vary according to the way the line from the centre of the sphere to a point is positioned), is shown in figure 7-15:

figure 7-15

The parametric equations are:

$$x = r\sin\phi\ \cos\theta$$
$$y = r\cos\phi\ \cos\theta$$
$$z = \cos\theta$$

Two sets of circular slices can be defined, and can be imagined from figure 7-4, in two ways. By keeping θ constant and varying ϕ, the set of slices have a common axis, with the circles all the same radius; they are the lines of longitude on a map of the earth. The lines of latitude correspond to slices of different radii, created by keeping ϕ constant and varying θ.

These circles are difficult to slot together because the ones with the same radii intersect in the centre of the sphere. This method of generating slices by keeping ϕ constant gives slices which are all on an axis is a property of many other parametric equations. The slices do not fit on a grid, the way we want them for Sliceform models. With one set of slices all intersecting on one axis, the models are rigid and they are harder to make, but not as Sliceforms. See the discussion of special ways of building models like this at the end of chapter 9.

To make a sphere from these parametric equations, they need to be converted to an explicit equation, which is easy to do by squaring and eliminating the functions of the angles, making use of the equation that for any angle θ.

$$\sin^2\theta + \cos^2\theta = 1$$

so eliminating ϕ from the equations for x and y gives

$$x^2 + y^2 = r^2\cos^2\theta$$

and then eliminating θ from this expression and the z parametric equation gives the equation:

$$x^2 + y^2 + z^2 = r^2$$

from which z can be determined easily if x and y are known. This is then an explicit equation and can be dealt with as described below. However, it is not always easy to convert such equations into explicit equations.

As the following examples of parametric equations show, not all such equations involve spherical polar coordinates. In all cases where an explicit equation cannot be derived, methods for plotting the slices require sophisticated use of computer programs as described in chapter 10, "Using the computer".

Examples of surfaces with parametric equations

Parametric equations are usually written with two variables u and v, to denote the angles. The examples in this section use this convention of using the parametric variables u and v.

Parametrically defined surfaces are often the easiest type of equation to appreciate how the variables affect the shape of the surface. These examples have been chosen, in part, to illustrate ways of designing models rather than just copying equations from books. They are not always suitable for making Sliceform models and, where they are not closed surfaces, some thought has to be given in making the model.

The following table shows the equation and suggested ranges to plot for the surfaces shown in the subsequent computer drawn illustrations and descriptions. The range of variables u and v is given in radians.

In the following diagrams of the surfaces, each one is shown as a stereoscopic pair, like figure 7-1. See the description under figure 7-1 for more details on viewing stereoscopic pairs.

"Cuboid sphere"

This surface (figure 7-16) is formed by modifying the parametric equation of a sphere (see above) by using the sines of all the sines and cosines in the equation; compare the two sets of equations to see what this means.

$$x = \sin(\sin(u)) \sin(\cos(v))$$
$$y = \sin(\cos(u)) \sin(\cos(v))$$
$$z = d \sin(\sin(v))$$

The height of the z value is also factored to make the surface symmetrical like a cube with d having the value 0.84147. The plotting ranges for u and v are $-\pi/2$ to $+\pi/2$.

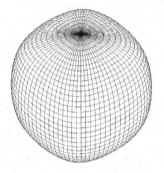
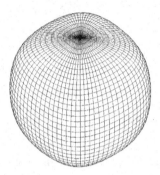

figure 7-16

If you compare the equations for this surface compared with the sphere, it is no longer easy to eliminate the parametric variables u and v and obtain an explicit equation. It is necessary to use the computer to define the slices.

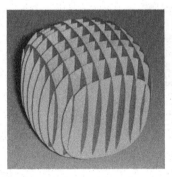
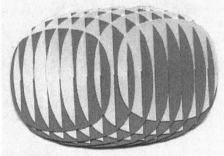

figure 7-17

Cross cap surface

This is an important surface in the history of mathematics since it is a model of the projective plane (see Fischer for more information) and so appears to be a kind of a simpler version of Steiner's Roman surface (see figure 4-3).

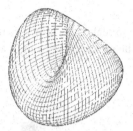

figure 7-18

The parametric equations are:

$$x = \cos(u)\sin(2v)$$
$$y = \sin(u)\sin(2v)$$
$$z = \cos^2(v) - (\cos(u)\sin(v))^2$$

They bear a superficial resemblance to the ones for the sphere and it can be thought of a sphere that somehow intersects itself in a line of singularity which is an axis of symmetry of two perpendicular planes. When making the slices, this then acts as the definition of the two sets of perpendicular slices, so that complete slices are obtained. The pattern formed by the flattened model is also quite attractive. The plotting ranges for u and v are 0 to π.

figure 7-19

Whitney umbrella

This is a very simple surface, having a singularity like the crosscap, but it is not closed.

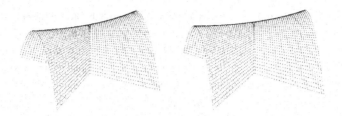

figure 7-20

The parametric equations are:

$$x = uv$$
$$y = u$$
$$z = v^2$$

This is simple enough for u and v to be eliminated so that the explicit equation

$$z = \frac{x^2}{y^2}$$

results, but this is much harder to plot as an explicit equation using computer software for two reasons. If you try to plot the equation as it stands, at the line of singularity there is division by zero which at one point divides zero by zero. If you rearrange it thus (assuming the software plots using z as the explicit function and so switching y and z in the parametric equations) then it becomes:

$$z = y\sqrt{x}$$

and invariably the plotting software cannot cope with the symmetry of z having two values because the square root has plus and minus values. This problem also arises if you plot the sphere from the explicit equation

$$z = \sqrt{r^2 - x^2 - y^2}$$

For actual plotting the equations must take the positive and negative cases and the two equations become

$$z = \pm y\sqrt{x}$$
$$z = \pm\sqrt{r^2 - x^2 - y^2}$$

So one should always be wary of the computer output and take note of the mathematics. So when making the models, you may only be able to plot half a slice and join the two halves in drawing the slice for the model. It all depends on the plotting software you are using.

Making a Sliceform model of the Whitney umbrella needs careful thought because of the line of singularity. Consider first making it as a solid surface (turned upside down as in figure 7-20 and then filled.) Only a slice which is central through the line of singularity would hold the model together. Slicing in other directions would give a number of slices which split into two parts.

The other possibility would be to make it as a recess in a solid block (again a turned upside version of figure 7-20) and this is the model shown in figure 7-21. The plotting ranges for u and v are -1 to 1 radians.

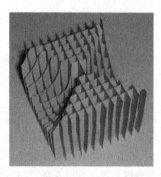

figure 7-21

Steinbach spiral

This is again a very simple equation, and a very elegant surface. Unfortunately, unsuitable for making a Sliceform model because of the many indentations.

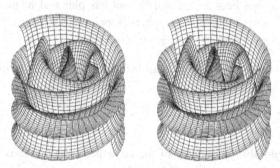

figure 7-22

This has been plotted with ranges for u and v of -2π to $+2\pi$. The parametric equations are:

$$x = u\cos(v)$$
$$y = u\sin(v)$$
$$z = v\cos(u)$$

Catenoid surface

This is an important surface in mathematics. It belongs to a set of surfaces called minimal surfaces. They are frequently met as soap bubbles. Soap bubbles are minimal surfaces because the forces acting on them go to a minimal state of equilibrium. This is why a blown bubble goes to a sphere, where the forces are not only equal, but they form the surface which has the least area for a given volume, namely a sphere. The catenoid surface is formed when two circular rings are placed in a soap solution and then withdrawn to give a film between the rings. When they held directly above one another the film forms a surface like a cylinder with a waist (very much like a hyperboloid of one sheet). However, because of the negative curvature, it is hard to make a good model without slices which are separated pieces, like the hyperboloid shown in figure 5-19.

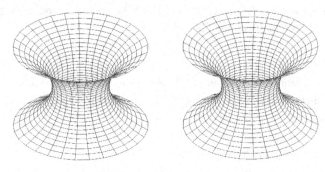

figure 7-23

This surface is also a surface of revolution, of a curve called a catenary. Not surprisingly, the catenary belongs to the set of minimal curves. If you take a chain and hang it between two points then the chain hangs in a catenary. Some suspension bridges are constructed as catenaries because of the property of minimisation of forces. The equation of a catenary is:

$$y = \cosh(x)$$

If we rotate the in space curve so that we plot $z = \cosh(x)$ instead, then it looks like the curve at the left of figure 7-24.

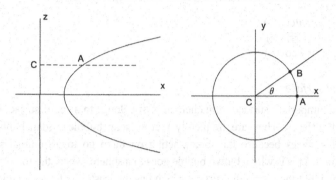

figure 7-24

Now suppose we have a point on the catenary, A, and we want to rotate it about the z axis. In the left diagram of figure 7-24, the dashed line represents a plane through A which is parallel to the *xy* plane. Now imagine you are looking along the z axis so that you see the line CA as the diagram on the right of figure 7-24 which is on this plane. Point C is the centre of a circle through A with CA its radius. If the radius of the circle equals a and the point A is rotated to B by an angle θ, the coordinates of point B are:

$$x = a \cos \theta$$
$$y = a \sin \theta$$

If we now redefine the equation of the catenary as $a = \cosh(v)$ where $z = v$ and combine these equations, also using u for the rotation angle, then we get the parametric equation of the catenoid surface:

$$x = \cos(u)\cosh(v)$$
$$y = \sin(u)\cosh(v)$$
$$z = v$$

The plotted version in figure 7-23 has a range for u of 0 to -2π and for v of -15 to 1.5 radians.

So if you want to convert any curve with an explicit function $z = f(x)$ to a parametric equation which defines the surface of revolution of that curve, then the same technique can be used.

Chapter 5, "Surfaces of revolution" describes how to make Sliceforms geometrically. Plotting them on the computer as slices on an xy grid is described in chapter 10, "Using the computer" in the section "Generating surfaces of revolution". A different approach is given later in this chapter in the section on "Sculpting with equations".

Cardioid based surface

This surface illustrates how a surface may be designed from a curve whose polar equation is known.

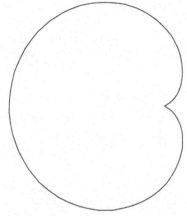

figure 7-25

The cardioid shown in figure 7-25 has the polar equation

$$r = a(1 - \cos \theta)$$

where *a* is a constant which determines the size of the curve, like the radius of a circle. Converted to a plane curve in parametric cartesian coordinates gives

$$x = a(1 - \cos\theta)\cos\theta$$
$$y = a(1 - \cos\theta)\sin\theta$$

Adding a third coordinate $z = u$ and changing the θ to v would give a parametric plot which is a cardioid shaped cylinder, since *x* and *y* only depend on *v*. But making the constant a into a variable value allows the cylinder to be restricted. Using sin(u) so that the value varies with the height *z* (and with $z = u$) and further restricting the values of *u* from 0 to π causes $\sin(u)$ to go from 0 to 1 and back to 0. The parametric equations are now

$$x = \sin(u)\,(1 + \cos(v))\cos(v)$$
$$y = \sin(u)\,(1 + \cos(v))\sin(v)$$
$$z = u$$

and the resulting surface (with plotting ranges of 0 to π for *u* and $-\pi/2$ to $+\pi/2$ for *v*) is shown in figure 7-26.

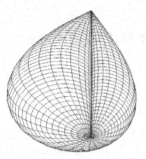
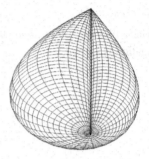

figure 7-26

Note that this is not a surface of revolution like the catenoid surface, but rather an extrusion with constriction. The model for figure 7-26 is shown in figure 7-27.

figure 7-27

Hyperbolic paraboloid

This surface is a quadric surface which does not have circular sections. The parametric equations are:

$$x = u$$
$$y = v$$
$$z = uv$$

This can be converted to the explicit equation:

$$z = xy$$

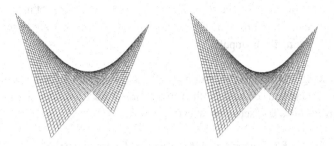

figure 7-28

Plotting with ranges for u and v from -1 to +1 gives figure 7-28. The lines forming the surface mesh are straight lines and this surface is a ruled surface. The model is described as such in chapter 6, "Ruled Surfaces".

Sliceform models from explicit equations

Sliceforms created from explicit equations are the easiest to plot and indeed the slices are the only ones that can be plotted without using the special techniques described in chapter 10, "Using the computer". In many ways they are the easiest of all Sliceforms to make because you can just plot the slices.

As mentioned above, explicit equations are of the form:

$$z = f(x,y)$$

where the function of x and y can take many forms. Then creating the slices is simply a matter of finding values of z for planes where first x values, and then y ones, are constant.

There are many ways you can find suitable surfaces for construction of Sliceform models. The obvious way is to look in mathematics books, or search the Internet. When you find the information, make sure you note the limits for plotting since such surfaces are not usually closed and some parts of the surface may not be very interesting. You should also be wary of surfaces that intersect themselves. Remember that you can always scale a model in a particular direction to make it easy to construct.

You can also invent your own surfaces, by plotting equations derived from families of two dimensional curves or by taking equations of known surfaces and varying them. It helps if you have a computer program which can plot mesh curves, because you can see if there are going to be problems by plotting the surface first. You can also see if the surface is going to be interesting enough to use in a model. Programs available to do this on the Internet are described in chapter 10, "Using the Computer".

The following descriptions of creating surfaces from explicit equations shows a number of standard equations, then there are further sections which show variations methods for "sculpting with equations".

Examples of surfaces with explicit equations

The following table shows the equation and suggested ranges to plot for the surfaces shown in the computer drawn illustrations in figures 7-29 to 7-34 These are fairly standard explicit surfaces. The values of z denote the maximum height of slices and so the height the box in which the slices will fit.

The following surfaces, apart from figure 7-34 which is a conoid, have no specific names. Figure 7-1 is a surface with an explicit equation called a monkey saddle surface; note how it is very similar to surface 3. The monkey saddle is the example used to show how to create these surfaces. The Sliceform model shown in figure 7-37.

These equations can be used to generate more than one Sliceform model. See the section "Multiple Sliceforms from one equation" later in this chapter.

In the following diagrams of the surfaces, each one is shown as a stereoscopic pair, like figure 7-1. See the description under figure 7-1 for more details on viewing stereoscopic pairs.

Figure 7-29 has the equation

$$z = \frac{6y}{x^2 + y^2 + 1}$$

and is plotted for a range of -3 to +3 for x, y and z.

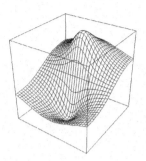
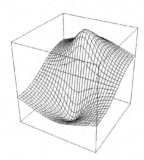

figure 7-29

Figure 7-30 has the equation

$$z = \frac{3\cos(x + y)}{x^2 + y^2 + 1}$$

and is plotted for a range of -3 to +3 for x, y and z.

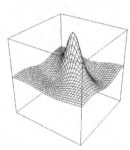
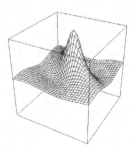

figure 7-30

Figure 7-31 with the equation

$$z = x - \frac{xy^2}{4}$$

is plotted for a range of -3 to +3 for x, and y and a range of -5 to +4 for z.

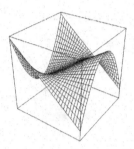 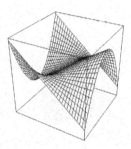

figure 7-31

Figure 7-32 has the equation

$$z = \frac{3\cos(x)}{2} + \frac{3\sin(y)}{2}$$

and is plotted for a range of -4 to +4 for x, y and z.

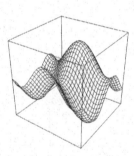 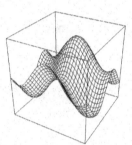

figure 7-32

Figure 7-33 which has the equation

$$z = \sin(xy)$$

is plotted for a range of -2 to +2 for x, and y and a range of -2 to +1 for z.

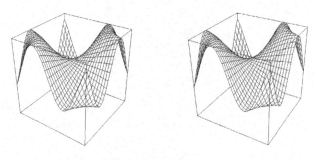

figure 7-33

Figure 7-34 is a ruled surface called the conoid, and its construction by a different method is described in the chapter on ruled surfaces, chapter 6. It has the equation

$$z = 0.05(xy^2 + x)$$

and is plotted for a range of 0 to 8 for x, -4 to +4 for y and a range of 0 to +6.8 for z.

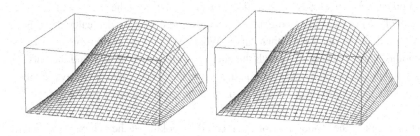

figure 7-34

The examples above are standard ones found in books describing equations for surfaces, usually because they were studied for a reason at the end of the nineteenth century. They are easy to make as Sliceform models since they are easy to plot. Other examples are given later in the chapter. Some of these are shown as Sliceforms in figure 7-35.

figure 7-35

Example, constructing a Monkey Saddle

As an example of creating a surface from an explicit equation as a Sliceform model, the following description shows how to create a Monkey Saddle surface. This surface has the equation:

$$z = a(x^3 - 3xy^2)$$

where a is a constant.

You can plot the curves of each slice by hand (calculating the points with a pocket calculator) or by using a computer. You need more points when calculating using a computer to get a smoother curve. When you plot by hand, you need fewer points since it is easy to smooth the curve as you draw it. If you find the curve is changing shape too quickly for you to smooth the points, you can always calculate intermediate points. More details on using a computer are given in chapter 10.

In designing the model, part of the method is looking at the symmetry to decide how many slices need to be drawn and deciding on the range. Often the hardest part is to know the range of coordinates to plot. With a closed curve like a sphere, the equation defines the range to be plotted, with surfaces like the ones plotted above the range is matter of judgement. A computer with a program that allows you to plot surfaces is useful, so that you can adjust variables in the equation to fit the size and shape of model you want to make and find the most interesting part of the surface. The following example of the Monkey Saddle surface has been chosen to point of the types of decisions that you may need to make.

1. Start with the constant a in the equation equal to 1. Plot the curve in a 3D plot program varying the plotting range values to select the best one.
2. In this case, a range of both x and y from -1 to +1 seems reasonable. This gives a maximum value for z of +2 and a minimum of -2. Taking the value of a as 0.5 means that the resulting surface fits in a cube.

3. Next, to decide how many slices need to be drawn, two considerations need to be taken into account. Firstly, from drawing a mesh picture as in figure 7-1, the surface is seen to be symmetrical for positive and negative values of y (because y^2 appears in the equation), but there is no symmetry in the other direction. This means that all slices for the various values of x need to be drawn but only the central slice and those on one side of the yz plane in the in the y direction.

4. The saddle point is where the surface is flat in one direction and in the other has a point of inflexion (see the slice in figure 7-36) where the curvature is different on either side of this point. It is important to include this as a point on both sets of slices in order that it is clearly visible. It occurs at the origin. The slice in the y direction (in the xz plane) has an edge which follows the z axis so it is a rectangle (see the model in figure 7-37). Slices with negative values of x have the opposite convexity to those where x is positive.

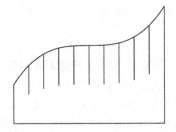

figure 7-36

1. These reasons dictate that there should be an odd number of slices in the range. The actual number chosen decides the smoothness of the surface.

2. The values of z range from -1 to +1 and there is no problem with adding any depth to the slices to solve the problem described earlier in the chapter and shown in figures 7-2 and 7-3.

The two views of the model shown in figure 7-37 use 18 slices, but a good idea of the surface can be obtained with six slices.

figure 7-37

Most models of surfaces from equations do not display an attractive pattern when the model is flattened. I tend to use the same colour card for both directions.

Having created this Sliceform Monkey Saddle to fit in a cube, it bisects the volume of the cube so that two models placed together will fill a cube because of its symmetry.

Multiple Sliceforms from one equation

Many surfaces are infinite and the parts that are used to create Sliceforms are only part of the surface. So one way in which one equation can be used to create different models is to make the plane forming the slice extend on different sides of the curve for each of the models. This can yield Sliceforms that look totally different, even though the surface being modelled is the same. This is not always the case, and the Monkey Saddle described above is such an exception because of its symmetry.

The Cassinian surface

In chapter 3, contours are used to define surfaces and examples are given of surfaces which are defined as families of curves. Although the following family of curves could be constructed using the contour methods, it is easier to work with equations.

Figure 3-7 shows a common way to draw an ellipse using the property that, for any point on the ellipse, the sum of the distances to the foci is a constant. Cassinian curves have the property that for any point, the *product* of the distances to the foci is a constant. The family of Cassinian curves is shown in figure 7-38.

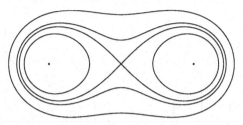

figure 7-38

The family has two "branches". Intermediate between the families is a figure-of-eight shaped curve called a lemniscate..The properties of this particular lemniscate were described by Bernouilli in 1694, and he called it after the Latin *lemniscus*, "a pendant ribbon fastened to a victor's garland". Outside the lemniscate, a member of the family is like an oblong curve with two dimples; but note that it is a continuous curve. Inside the lemniscate, the curve breaks up into two parts, sometimes called Cassinian ovals because of their shape.

The equation of the Cassinian curves is given by:

$$[(x - a)^2 + y^2][(x + a)^2 + y^2] = k^4$$

where the foci have coordinates $(-a,0)$ and $(a,0)$ and the product of the distance of any point from the two foci is k^2.

To create the Cassinian surface from this equation, replace k by z

$$[(x - a)^2 + y^2][(x + a)^2 + y^2] = z^4$$

Now the various Cartesian ovals are the "contour" lines and to create a Sliceform model use the same technique as for the Monkey Saddle surface above for differing values of x and y:

1. Plot one set of curves for the slices keeping x constant.
2. Plot the other set keeping y constant.
3. Ensure that there are slices which correspond to x being zero in the first set and y zero in the second set. Otherwise you will miss the important parts of the surface, namely the tips of the two peaks and the bottom of the valley between them. See the section "The number and position of slices" in chapter 1.

Now because taking the root of z^4 can give two values, a positive and negative, two curves can be drawn to yield slices for two models (a "mountain" type and a "valley") type. The central slices for these, corresponding to the horizontal direction through the foci of figure 7-38 are shown in figure 7-39.

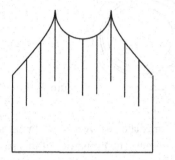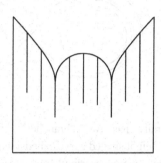

figure 7-39

The two Sliceforms are shown in figure 7-40 and can be fitted together to form a cuboid.

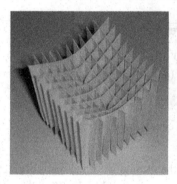

figure 40

Other suggestions for variations to explore are:

- The positive and negative curves could be combined to make slices for a different surface with a valley on one side and a mountain on the other
- Each slice could be reflected about a horizontal axis to create a symmetrical solid
- z could be equated to k^2 or k^4 instead of k, to give three more surfaces which could be subjected to the same combination techniques.

When a model is made for z equated to k^2, it is much higher than the corresponding one in figure 7-40.

Confocal ellipses and hyperbolae

Figure 3-8 shows how a set of ellipses with the same focal points can be used to define a contour. Although the methods of chapter 3 could be used to create the

Sliceform model, the equation method is easier, not least because it can be pre-viewed on a computer. This confirms the contour image of a very steep surface. Models which are very high compared with their width do not make very good models as described at the end of chapter 1. However, all is not lost since the concept can be generalised to include confocal hyperbolae and this does lead to a reasonable model.

Ellipses and hyperbolae are known as central conics, since they have two axes of symmetry. The equation of a central conic is:

$$\frac{x^2}{a^2} \pm \frac{y^2}{b^2} = 1$$

where the constants a and b determine the shape of the conic with the plus sign giving an ellipse and the minus sign a hyperbola. Suppose the parameter z is introduced into the equation as follows:

$$\frac{x^2}{a^2 - z} \pm \frac{y^2}{b^2 - z} = 1$$

then the equation defines a family of confocal conics where each value of z defines one conic. In a moment z is going to be the height of the surface, but for now think of it as defining a particular conic (not just an ellipse or hyperbola). For convenience, let a be greater than b and consider the plus sign. If z is less than b^2 the conic is an ellipse. If z is equal to b^2 the conic consists of the x axis. When z is greater than b^2, the conic becomes a hyperbola. For full details consult a book on the algebraic geometry of conics. Figure 7-41 shows both confocal ellipses and hyperbolae in the same diagram.

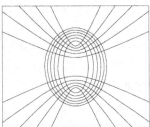

figure 7-41

If we now think of this equation representing a surface and we solve for z so that we can plot the slices, the equation reduces to a quadratic:

$$z^2 + dz + g = 0$$

where $d = x^2 + y^2 - a^2 - b^2$ and $g = a^2b^2 - a^2y^2 - b^2x^2$ and the quadric can easily be solved for z with the usual formula to give:

$$z = \frac{-d \pm \sqrt{d^2 - 4g}}{2}$$

When the surface is plotted using the negative value of the square root, this gives the confocal ellipse surface (figure 7-42) which you can see is very thin and so not appropriate for making a model.

In both of the following surfaces, a has the value 1 and b the value 2. The range for plotting is -10 to + 10 for x and y.

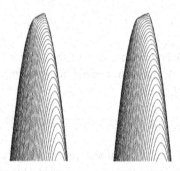

figure 7-42

Using the positive value of the square root, the confocal hyperbola surface (figure 7-43) which looks like a piece of pastry that has been gripped between thumb and finger and crimped:

figure 7-43

Plotting the two surfaces together on the same diagram is not easy, but they intersect, so there is potential for exploration. As with the Cassini surface, the Sliceform model can be made two ways, using indented versions of the figures shown here. The Sliceform of the confocal hyperbolae surface is shown in figure 7-44.

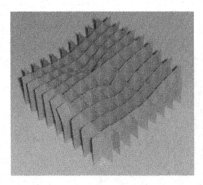

figure 7-44

Trousers surface

An interesting surface which shows two very different models comes from the equation

$$z = \sqrt{\frac{x\sin(x) + y\cos(x)}{2}}$$

Plotting ranges are -4.1 to +4.1 for x, -5 to +5 for y. Just plotting the positive root of the equation gives the result shown in figure 7-45 which also has the xy plane cutting through it. This is an artefact resulting from the way plotting software has to cope with square roots.

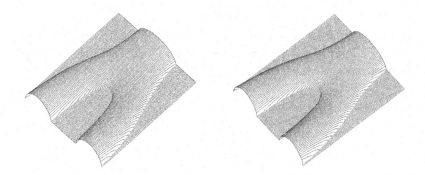

figure 7-45

This makes an interesting Sliceform model as it is. The model of the surface in figure 7-46 (with a base added) does not look as smooth as the computer version because there are more mesh lines in the computer one. By plotting each slice and including the positive and negative values for the square root, then a model which looks like a pair of trousers is easily obtained.

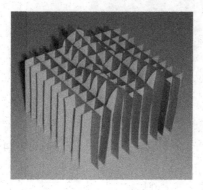

figure 7-46

Plotting using this explicit equation also illustrates the difference between plotting explicit and implicit equations. If the implicit equation is plotted by squaring the equation then the result of figure 7-47 shows the surface as a whole.

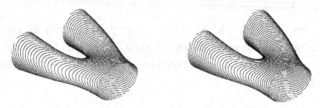

figure 7-47

The Sliceform model is very unusual because of the way it changes into two branches. It also looks very architectural (figure 7-48), although as a paper Sliceform model is not as rigid.

figure 7-48

The plot using the explicit equation can also be used to plot an indentation model, thus giving three possibilities for the same surface.

Slicing the torus

A torus is most easily described as a surface of revolution. Take a circle in the zx plane and rotate it about the z axis as shown in figure 7-49. This shows the three qualitatively different cases depending on the position of the circle relative to the axis of rotation. The axis either does not cut the circle, is tangent to it, or cuts it, giving the ring, horn or spindle torus respectively.

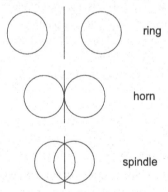

ring

horn

spindle

figure 7-49

The cutting away of the horn and spindle toruses in figure 7-50 show how they have an internal structure not present in the ring torus.

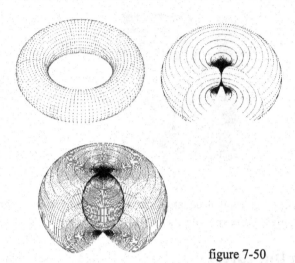

figure 7-50

The following slicing of the torus concentrates on the ring torus. Figure 4-6 shows slicing a ring torus to obtain a series of curves known as the spiric sections of Perseus. Considering the torus as a surface of revolution of a circle which lies on a plane through the axis of revolution, it is obvious that a torus can be sliced in to give two sets of circles. Firstly, a slice using a plane through the axis, for example the central slice of figure 4-6, is a pair of circles (the ones being rotated). Another set is obtained by slicing through a plane perpendicular to the axis which gives a circle as a rotation of each point of the defining circle. There are also other circles called the Villarceau circles. The model of the torus is built up of a set of spiric sections and a set parallel to the Villarceau circles.

Equation of the torus

The explicit equation of a torus is

$$(x^2 + y^2 + z^2 + a^2 - b^2)^2 = 4a^2(x^2 + y^2)$$

where a is the distance of the circle being rotated from the centre and b is its radius; the circle is rotated about the z axis. However, to create a Sliceform model, it is more convenient to work with a parametric equation and see how it is built up in stages. The circle is of radius b whose centre is a distance a from the axis and is in the zx plane and will be rotated about the z axis. Its equation is:

$$x = a + b\cos(v)$$
$$y = 0$$
$$z = b\sin(v)$$

A rotation of a point (x,y) about the z axis by an angle θ is given by

$$x_{new} = x\cos(\theta) - y\sin(\theta)$$
$$y_{new} = x\sin(\theta) + y\cos(\theta)$$

so using the parameter u for the rotation, the parametric equation of the torus becomes

$$x = (a + b\cos(v))\cos(u)$$
$$y = (a + b\cos(v))\sin(u)$$
$$z = b\sin(v)$$

We are now in a position to look at the slicing of the torus for creating the Sliceform model.

Making Villarceau's slices

Villarceau showed that if you slice a ring torus with tangent planes through the centre of the torus as in figure 7-51, then the section is a pair of intersecting circles as shown in figure 7-52.

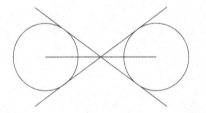   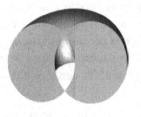

figure 7-51   figure 7-52

Slicing for the model

To make a model, to show both the spiric sections and so have the circles being rotated as a slice as well as the Villarceau circles, it is necessary to choose the radii of the torus and consider the tangent planes as in figure 7-53.

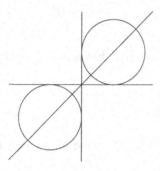

figure 7-53

This diagram shows the torus rotated about the y axis by 45° and a section through the zx plane. The tangent planes are the xy and yz planes. The radii of the circles defining the torus are $\sqrt{2}$ for a and 1 for b. Then if the slices are made parallel to the zx plane (keeping y constant) the sections are the spiric sections of Perseus and slices parallel to the yz plane (keeping x constant) the sections are ones including the Villarceau circles. To find the equation of this torus, it is necessary to rotate the parametric equation by an angle of 45° about the y axis using the same technique as for creating the equation of the torus from that of the circle. So the new equation becomes:

$$x = (a + b\cos(v))\cos(u)\cos 45° - b\sin(v)\sin 45°$$
$$y = (a + b\cos(v))\sin(u)$$
$$z = (a + b\cos(v))\cos(u)\sin 45° + b\sin(v)\cos 45°$$

There is more about the mathematics of the Villarceau circles (including a proof of why they are circles) in R J T Bell, Coxeter and Ronik.

The top row of figure 7-54 shows the Villarceau circles slice plus one adjacent slice and the bottom row shows two of the spiric section slices which because the torus is rotated, have slots are not along the axis of symmetry.

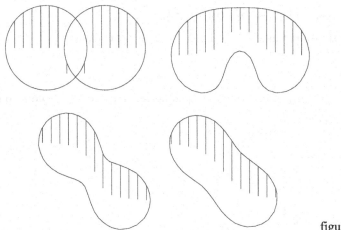

figure 7-54

The model is shown open and closed in figures 7-55 and 7-56. It has an added attraction because one set of slices are inclined at an angle and the flattened model also has a three dimensional quality because of this. You can see the Villarceau circles in figure 7-55 behind the hole in the torus with one of the circles from the spiric sections touching at the intersection points.

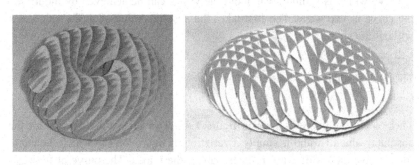

figure 7-55   figure 7-56

Sculpting with equations

It is possible to create surfaces of revolution (see chapter 5 "Surfaces of revolution") by using explicit equations. The method for doing this leads to other ideas for creating new surfaces by "sculpting" with these equations for the surfaces of revolution.

Curves to surfaces of revolution

The curve shown in figure 7-57 is known as an exponential curve. Its equation is:

$$y = e^{-x} \qquad\qquad \textit{equation A}$$

and has the property of being very steep at the left and being asymptotic to the x axis at the right.

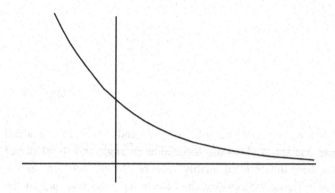

figure 7-57

A curve which is symmetrical about the y axis can be achieved by modifying the curve in two ways. The first takes the absolute value of x (the value of x without the sign):

$$y = e^{-abs(x)} \qquad\qquad \textit{equation B}$$

The second squares the value of x:

$$y = e^{-x^2} \qquad\qquad \textit{equation C}$$

These yield the curves shown in figure 7-58. The left hand curves is for the absolute value of x and is simply a variation of figure 7-57 in which the curve for positive values of x are reflected about the y axis. The curve at the right uses x^2 and is known as a bell curve. All these curves are met in statistics.

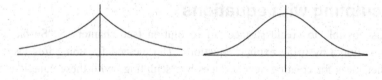

figure 7-58

To convert these curves to surfaces of revolution, they need to be considered as curves in a plane rotating about the z axis, with a slight change of equation. Figure 7-59 shows the bell curve rotated about its central vertical axis.

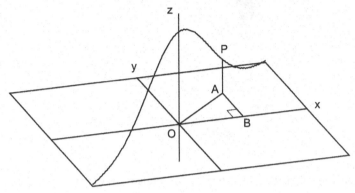

figure 7-59

Consider the point P of the curve and its projection onto the xy plane at point A and consider the plane in which the curve lies. Compare distances/coordinates with the ones used to plot the curve in the usual cartesian xy plane. The distance OA corresponds to the original x coordinate and the distance AP to the original y coordinate. So the equation becomes:

$$AP = e^{-OA^2} \qquad\qquad equation\ D$$

The height AP is the z coordinate of the curve in space, and OA can be obtained from Pythagoras's theorem:

$$OA^2 = x^2 + y^2 \qquad\qquad equation\ E$$

The equation defining the point AP is for one point on the curve in a particular plane. For the surface of rotation, the equation becomes generalised:

$$z = e^{-(x^2+y^2)} \qquad\qquad equation\ F$$

and the surface appears as in figure 7-60

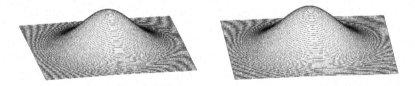

figure 7-60

The Sliceform version of this surface is shown in figure 7-61.

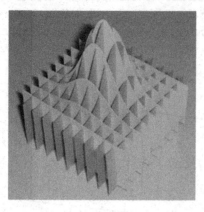

figure 7-61

The equation taking the absolute value the equation becomes:

$$z = e^{-b\sqrt{(x^2+y^2)}}$$
                                                              equation G

with a constant b added as an extra refinement to raise the height of the tip to give a more interesting surface. The absolute function is not needed as here if only the positive value of the square root is taken; the length equivalent to AB in figure 7-51 is always positive. The plotted surface in figure 7-54 has a value of 2 for the constant b.

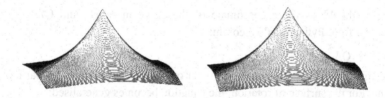

figure 7-62

The Sliceform model of the surface is shown in figure 7-63.

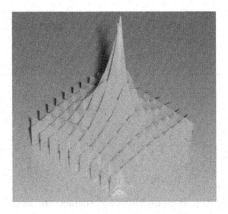

figure 7-63

Modifying the equation to make "craters"

Such equations make interesting Sliceforms, but they can be enhanced to yield even better Sliceforms. The following method creates surfaces which look like craters.

Figure 7-64 shows a modification of the curve of equation B to

$$y = e^{-b \cdot abs(x - d)} \qquad \text{equation } H$$

where the value of the constant d moves the curve along the x axis.

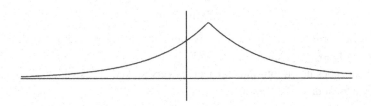

figure 7-64

if the absolute value of x is used as well (equation I) , then the curve is symmetrical about the y axis and has split into two peaks as shown in figure 7-65. The separation of the peaks is defined by the constant d, to correspond to the move introduced in equation H

$$y = e^{-b \cdot abs(abs(x) - d)} \qquad \text{equation } I$$

The curves in figure 7-6 show the effect of two different values of d and how a larger value gives a smoother curve between the two peaks.

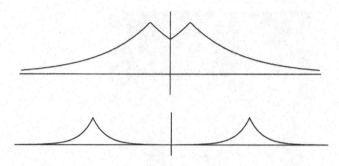

figure 7-65

Now converting equation *I* to a three dimensional surface and noting that only the positive value of the square root is taken gives:

$$z = e^{-b \cdot abs(\sqrt{(x^2+y^2)} - d)}$$ *equation J*

The surface when plotted looks like figure 7-66.

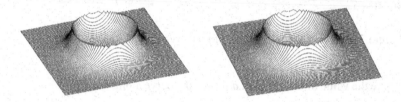

figure 7-66

The Sliceform model (figure 7-67) is more attractive because of the slices not being able to give a circularly smooth rim to the crater. The sharp edge of the rim as it goes down into the crater gives very spiked slices which almost give it the appearance of a flower.

figure 7-67

Similar working of the bell shape and an extension of equation G gives a crater which has a smoother edge to the rim. The surface equation is

$$z = e^{-b(\sqrt{(x^2+y^2)}-d)^2} \qquad \textit{equation K}$$

and the surface is shown in figure 7-68.

figure 7-68

In the Sliceform model (figure 7-69), as with the sharp-edged crater, the slices add an effect not evident in the computer plotted version, which is especially striking when the slices are different colours and the model is rotated.

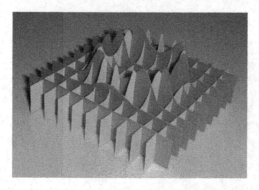

figure 7-69

Craters which are not surfaces of rotation

It is possible to be more creative in sculpting with the equations. The following is just one example.

By defining the exponential function using $\sqrt{x^2 + y^2}$ in equations J and I, this is equivalent to measuring radially from the origin. Using a constant d causes any movement from the origin to be constant in all directions. By replacing the constant d by a function of x and/or y, then the rim of the crater can be moved and the rotational symmetry destroyed. Such a function is the polar equation of a cardioid (figure 7-70):

$$r = a(1 - \cos \theta) \qquad\qquad equation\ L$$

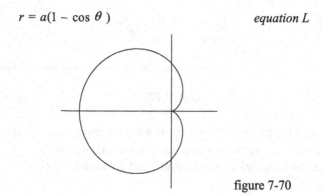

figure 7-70

In order to use this polar equation, it is necessary to calculate the angle for a point with coordinates (x,y) and then determine its cosine which is easily done by calculating as

$$\frac{x}{\sqrt{x^2 + y^2}}$$

so that d is determined from equation M.

$$d = a\left(1 - \frac{x}{\sqrt{x^2 + y^2}}\right)$$

equation M

Note that the origin is at the cusp, and the crater will be distorted from this point and the slope of the walls of the crater will also not be symmetrical as they were with equations J and K. In figure 7-71, you can see that the walls are very steep at the cusp at the right, but much gentler at the left.

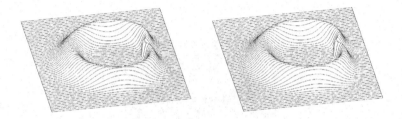

figure 7-71

Figure 7-71 is a plot of the surface defined by equation K, with d replaced by the expression in equation M. The values of a and b are 2. In the two views of the Sliceform model (figure 7-72) the cusp is not as pronounced as in the computer version, because of the number of slices used and the high walls. There is much scope for exploration on just this model, as indeed is the case with many of the models in this chapter.

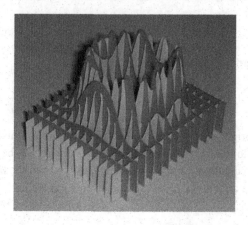

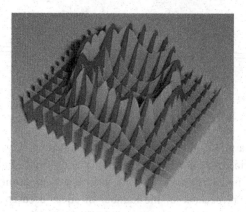

figure 7-72

Chapter 8
Polyhedra

Mathematicians often debate whether polyhedra are surfaces or solids. Whatever your view, creating Sliceform models from them allows exploration of many of their properties.

Slicing polyhedra

Many geometrical puzzles are based on making a slice through a polyhedron. The first two of the following examples are the most well-known ones.

The tetrahedron puzzle

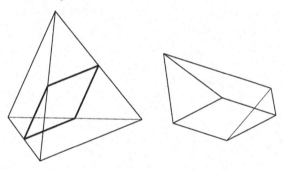

figure 8-1

This puzzle consists of two pieces which have to be assembled into a tetrahedron. They are made by slicing a tetrahedron to give a square cross-section. Many people have great difficulty in putting the two pieces together so that they are rotated and not mirror images of one another. This puzzle is the basis of the tetrahedron Sliceform.

Slicing a cube

Most models in this book can be considered as a cube that has been sculpted in some way, since the slices are made using a regular grid and the slices are perpendicular to one another. In chapter 2, the basic model was described using a cube. The slicing puzzle for the cube is to find a slice which is a regular hexagon. This is achieved by joining the mid-points of sides thus:

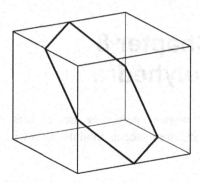

figure 8-2

There are other options for slicing in different directions, which do not have to be perpendicular.

Slicing an octahedron

The octahedron can also be sliced in many different directions. It too can be sliced to give a regular hexagon:

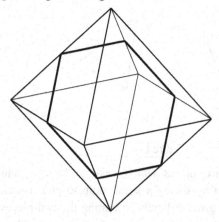

figure 8-3

The obvious way to slice an octahedron in half is to make two square pyramids by cutting along the edges. Making more slices parallel to the base of the square pyramid gives more squares. However, if you make equally spaced cuts for a Sliceform, the model does not fold well. Overcoming this problem is described later in the chapter.

The Dodecahedron and Icosahedron

The remaining two Platonic solids can also be sliced to give a regular polygon. Both give a regular decagon. The dodecahedron can be sliced to give a regular hexagon and the icosahedron an irregular dodecagon (with equal length sides and unequal angles). See Holden for pictures of models of these and other polyhedral slicing.

Designing Sliceform Polyhedra

Some polyhedra are easy to slice since it is easy to see slices as perpendicular sets of planes in the two directions. Perpendicular sets of planes also make it easy to determine how the slots should be cut. In some cases you may be able to see the shape of the slices, but how the slots should be cut may not be as obvious.

Polyhedra offer many possibilities for slicing in different ways, and are well worth exploring to find out how they fold flat in a variety of patterns.

General tips on designing

In designing Sliceform polyhedra, the fundamental things you need to identify are, what are the shapes of the slices and how do you draw the slots.

Although you may have an idea of how you wish to slice, as many of the examples in this chapter show, picturing the slices and how their shape varies requires some sort of sketch for most people. A two dimensional sketch may help, but a better method is to make a net of the polyhedron from paper, fold it up and draw lines on it with a felt tip pen. Use this to make a better version, with lines drawn accurately on another net. You may need to use different nets and models to look at the slices in different directions.

One problem that arises frequently is that you end up with insufficient slots in the end slices. Combining the markings for the different directions (preferably using different coloured pens) will help overcome this. It will also help you to decide how the slots are placed.

The position of the slots is a much harder idea to visualise. Usually the central slot is easy to work out. The slots are parallel, so moving them to the other slices is then a matter of locating an equivalent position on each one of the slices.

Sliceform tetrahedra

The tetrahedron puzzle at the beginning of this chapter leads naturally to a Sliceform. Varying the number of slices can give rise to models which show different effects. When I teach Sliceforms, I use a simple model with six slices (figure 8-4a) since it is quickly assembled and is partly a puzzle. The more advanced model with eighteen slices (figure 8-4b) is then easier to make.

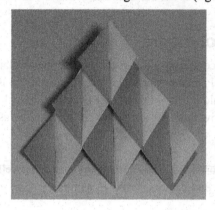 

figures 8-4a and 8-4b

The flattened version of the model with six slices is shown in figure 9-17.

Templates for making your own tetrahedra are given the appendix at the end of the book.

A net for the tetrahedron puzzle pieces

I have made the simple tetrahedron Sliceform model at many workshops and though it is easy to cut out, it is surprising how some people take a long time to see the rotational symmetry. So I often use the puzzle at the beginning of the chapter to explain what is happening with the following net of a tetrahedron cut in half to understand the design method and how the slices fit together.

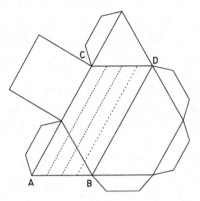

figure 8-5

Make two of them and put them together to make a tetrahedron by fitting the squares together. The dotted lines show the edges of the rectangles that make up the slices. Note how, as the rectangle moves to the edge from being a square to a straight line, the perimeter of the rectangle remains constant. You can see why this is by the lines which form two sides of the rectangle in the parallelogram ABCD in figure 8-5

Having designed the shape of the slices, the slots have to be added. Imagine a tetrahedron in a cube like figure 8-6

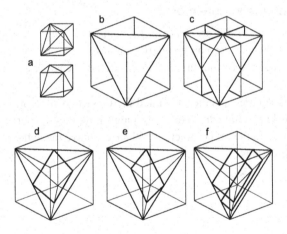

figure 8-6

Figure 8-6a shows how a tetrahedron fits in a cube. Figure 8-6b shows the cube turned around and 8-6c where two central slicing planes have been placed in the cube. In figure 8-6d, one of these planes cuts the tetrahedron in half to give a square (like figure 8-1). In figure 8-6d another square is formed where the other plane of figure 8-6c cuts the tetrahedron, so the two square slices of the tetrahedron are at right angles to one another. These two will obviously fit together so that their vertical diagonals slot into one another. This means that the slots are at an angle of 45° to the edge of the square. So when another slice is added as in the case in figure 8-6f, this too must have slots which are at 45° to the edge. You will see this much more easily of you make the model. For a simple version of the model, made up of six slices, the slices for one direction are only of two shapes, one square and two identical rectangles and look like this.

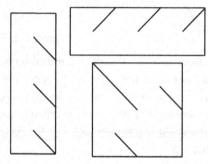

figure 8-7

Note that the slots in the square are not all in the same direction. This helps to hold the model together as with the hyperbolic paraboloid in chapter 6.

Assembling the simple model

The slices for figure 8-7 are given in the templates. The model only has six slices and so should take you about 10 minutes to cut out and assemble. Cut out all the pieces and cut out the slots as described in chapter 2.

Fit the centre squares together, then add the other two slices in one direction. Use the two half tetrahedra models to judge how they are orientated. Then insert the slices in the other direction. When you have made the model, fold it flat in two directions.

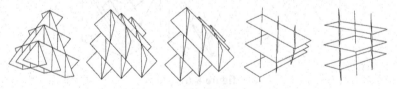

Figure 8.8

Assembling the advanced model

The template appendix, has a set of slices for a more complex tetrahedron with nine slices in each direction. In this model, all the slots are in the same direction in each slice. This model has many more pieces and takes up to an hour to make. With more slices the tetrahedron looks more solid. Make the simple model first so that you can see the structure. Take extra care not to cut the slots too wide or you will find that the pieces fall out in the early stages. Cut out the central squares and their slots and fit them together. Then cut the next four slices (two in each direction) and add them to your model. Continue cutting and adding four slices at time until you reach the outside. If you make the model with two colours of card it will not only be a more interesting model, it will also be easier to make.

When you have made the model, fold it flat in two directions. With it folded, tap the long flat ends gently on a flat surface to distribute the slices evenly in the slots.

Sliceform cubes

The cube is the simplest polyhedron for making a Sliceform, and indeed many of the other models are derived from it. In its simplest form, all slices are squares in both directions and so only one slice needs to be drawn for the template (figure 8-9).

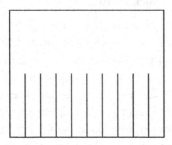

figure 8-9

While, simple to make, this Sliceform does not have any special properties. When it is folded flat in both directions, then the shape is the same, a rectangle. I leave it to the reader to work out its ratio, only saying that it is *not* $1:\sqrt{2}$.

As shown at the start of the chapter, a cube can be sliced to give other polygons and this does give more interesting models as well as stimulating thought as to the shapes of the polygons as the slice positions vary. The two directions for the slices also add more variations.

Slicing a cube diagonally

Making a slice using two diagonally opposite edges of the cube gives a rectangle with sides in the ratio 1:√2. Making a set of parallel slices as you move to the other pair of parallel edges gives rectangles which collapse to a line when you reach the edges. The model could then be constructed as a set of rectangles whose height is always the height of the cube.

Making identical slices in each direction

However, making a regular division causes problems, which arises in the model of the octahedron later in the chapter. It is illustrated by drawing the grid and then looking at the end slice (figure 8-10).

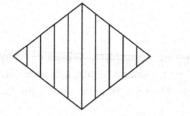 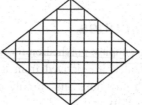

figure 8-10

In the diagram on the left, with one set of slices shown from above, all looks well, but when the second set are added as shown in the middle, two problems are apparent. Firstly, the ends of the slices in each direction touch tip to tip. This shows up more when you try to flatten the model as the slices do not want to move smoothly. Secondly, as shown on the right, the slices at the end only have one slot and are not supported enough. There is not a unique force acting on the slice when you try to flatten the model; the slice is "floppy" and its position in the opened model is not defined. Moreover, with only one slot, there is not enough friction to hold the slice and it falls out easily.

The solution to this is to make the spacing of the slices irregular. This gives a set of diagrams equivalent to figure 8-10 as shown in figure 8-11.

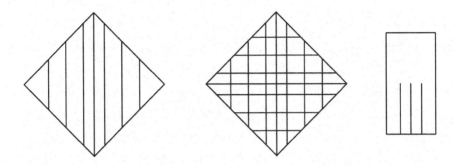

figure 8-11

Note that in the central diagram, the positions of the slices no longer meet at the square perimeter and also that the end slot as at the right now has three slots.

The complete model is shown in figure 8-12.

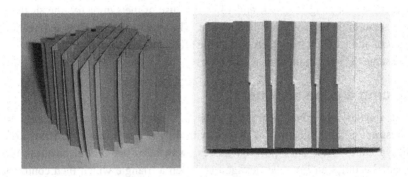

figure 8-12

This time the rectangle formed when the model is folded flat does have the ratio 1:√2. Note the pattern of stripes reflecting the irregular slicing.

Making different slices in each direction

The regular line pattern in the left of figure 10 can be used to make a model if the other set of slices are parallel to a side of the cube, giving squares.

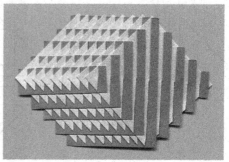

figure 8-13

One set of slices then consists of rectangles with one side equal to the line pattern in the left of figure 8-11 and the other the side of the cube. The other set are identical square slices. Then the squares are cut parallel to a diagonal, along the position of the lines in the in the left diagram of figure 8-11.

As you can see in figure 8-13, the model flattens to a hexagonal shape with equal sides but angles of 90° and 135°. Because of the symmetry, the shape is the same both ways when it is flattened. The spacing of the slots defines the pattern and explains why there is a stripe effect on one side. The spacing is not the same distance in the two sets of slices.

Slicing a cube obliquely

The regular hexagonal slice of a cube in figure 8-2 can also be used to construct a cube Sliceform, although with only a few slices it may not look like a cube. If the slice position moves away from the central position, the shape becomes an irregular hexagon and then a triangle which then contracts to a point at the vertex of the cube. All the triangles are equilateral and all the angles of the irregular hexagons are 120°.

The construction of the slices presents much more of a challenge than the polygonal slices described so far. Although the cube is there as framework, the slices are not parallel to its sides and they are not at right angles to one another either. If this were not enough, when constructing the grid, it is very easy to get into the situation of having slices which have only one slot as with the example in figure 8-10. There is an element of symmetry helping out, with the shapes of the slices, but the spacing of the slots also has its own symmetry.

The first step is visualisation of the slices. In the following set of cases, the middle regular hexagon is also shown. As the slicing plane moves to the corner of the cube, the slice is a hexagon with two different length sides. Then as in the centre of figure 8-14, three alternate sides of the hexagon have contracted to a point which is also a vertex of the cube and then slices are a triangle.

figure 8-14

Slicing on the other side of the central regular hexagon, gives a similar set of slices with the triangles pointing to the top left corner of the cube.

The next stage in the design is to draw the grid. This is visible by looking along the plane of the central hexagons. The cube drawn in parallel projection appears as a rectangle of the same shape as the diagonal plane of the cube, that is with a ratio of 1:√2. The near edge of the cube bisects the rectangle.

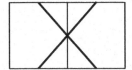 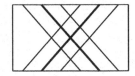 

figure 8-15

Since the vertices of the central hexagon are half-way along the edges of the cube, they appear a quarter the way along the edges of the rectangle as in the left diagram of figure 8-15. The other slices are positioned in parallel planes which appear as parallel lines in the other two diagrams. The central one shows some irregular hexagon positions and the one on the right the special triangles at the corners of the cube.

This is sufficient to create a simple model, with six slices. Constructing a model with more slices has all the problems of the other method of slicing the cube diagonally as described above, such as slices not having enough slots, together with positioning the slots. I leave it as a challenge to the reader. The six slice model is shown in figure 8-16.

figure 8-16

Figure 8-17 extends the right hand drawing of figure 8-15 to show how the length of the sides of the slices is determined. The lines of the diagram from figure 8-15 define the grid and so the position of the slots on the mid lines of the hexagonal and triangular slices. The square at the top in the left hand drawing of figure 8-17 is the square of the cube, that is the plan view of the cube. Looking at the central drawing of figure 8-15, the diagonal of this square is the edge of the triangular slice. The other line marked on the square is the side of the hexagon.

The two hexagons and four triangles each have the same dimensions because of the symmetry. The shapes of the slices with slots marked are shown at the right.

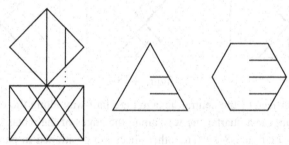

figure 8-17

Sliceform octahedra

As with the cube, there are many ways to slice octahedra. Also, since the octahedron can be thought of as a pair of square pyramids, fitting base to base, any Sliceforms designed from an octahedron can be cut down to make another model based on a pyramid.

Slicing an octahedron parallel to the axes

The obvious way to slice an octahedron is to make a series of slices which are square, the squares being parallel to what would be the base of a square pyramid.

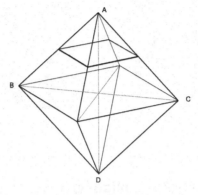

figure 8-18

Slices parallel to the plane through pairs of axes, are all squares. Such a slice is shown in figure 8-18. The other set of slices would then be parallel to another plane through a pair of axes such as the plane ABCD.

However, equally spaced slots would give the same problem as described above for slicing a cube diagonally. The solution is to use the method in figures 8-10 and 8-11, and even to use the same diagram in another model. For a model with seven slices in each direction, the different types of slice are:

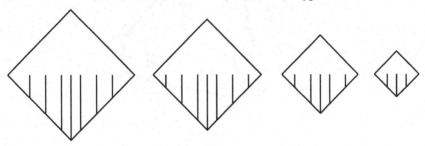

figure 8-19

Compare the spacing of the slots with the spacing used in figure 8-12.

The Sliceform model of the octahedron made from these slices has an interesting pattern when folded flat.

figure 8-20

Slicing an octahedron obliquely

As with the cube, the octahedron can be sliced obliquely to give a cross-section which is a regular hexagon (see figure 8-3), by joining mid points of selected edges. If the slicing plane moves parallel to this plane, the hexagon becomes irregular, with alternate edges having the same length.

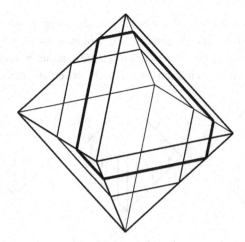

figure 8-21

Three sides of the hexagon contract and the other three sides expand, so that the slice eventually becomes one of the triangles of the octahedron. The edge lengths of the hexagons are easier to determine than in the equivalent cube case in figure 8-16. The edge lengths are measured on the surface of the octahedron,. Two of the edge lengths are shown in the following diagram.

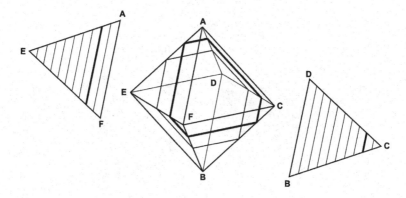

figure 8-22

The spacing of the slices can also be made equal, so the positions of the slots are also easier to find. They are shown in figure 8-24.

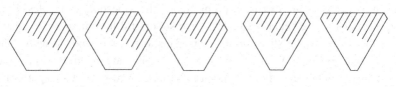

figure 8-23

This gives the model in figure 8-24:

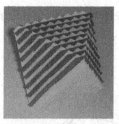

figure 8-24

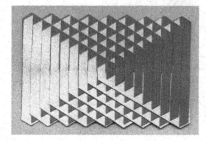
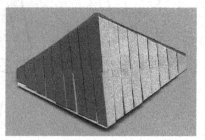

Sliceform prisms

Prisms are formed when a polygon is "extruded", that is moved in a direction perpendicular to the plane in which it is drawn. They are easily converted to Sliceforms. First draw a grid using lines parallel to two sides of the polygon. Then, because of the extrusion property, all slices become rectangles. Even sided polygons are better because the two set of grid lines intersect to ensure that there are enough slices on the slots. This avoids the problem with slices only having one slot. With odd sided polygons, drawing the grid lines parallel to sides of the polygon means that slices become narrower as they approach a corner. This results in slices with only one slot arising quite easily.

The following grids on a regular hexagon and an equilateral triangle show the difference between using even and odd polygons.

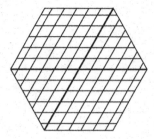
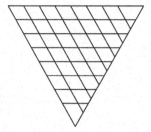

figure 8-25

In the grid for making a hexagonal prism on the left, the central slices are marked in bold. Note how the other slices do not slot in centrally. This makes it an interesting puzzle during assembly. When flattened, the width of the resultant rectangles differs markedly. The pattern of stripes is interesting because it is not uniform as is the case with a square prism (that is a cube).

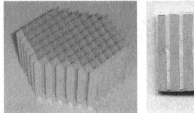 

figure 8-26

A Sliceform rhombic dodecahedron

This is another model where drawing the edges of the slices on a paper model and looking for symmetry relationships shows that very few slices need to be designed. By using symmetry, when the model is flattened in both directions it gives the same result.

A rhombic dodecahedron is one of the Catalan solids which are duals of the semi-regular solids (the Archimedean solids). It is the dual of the cuboctahedron. All its faces are the same shaped rhombus which has diagonals in the ratio of 1:√2. Seen in the orientation as shown in figure 8-27, there is a band of four rhombi with their longest diagonal bounding the "equator". They fall into two pairs: the pair with diagonals AB and CD and the pair with diagonals BC and AD. Each pair is parallel to one another. Each rhombus of a pair is perpendicular to the rhombi of the other pair in a plane, that is the diagonals AB, BC, CD and AD are successive sides of a square.

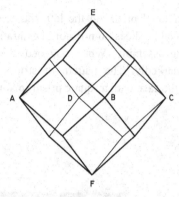

figure 8-27

This symmetry means that only one set of slices need to be designed. In figure 8-27, some of these are constructed as follows. The slices are parallel to the face with diagonal BC and as the slicing is moved from the centre, the vertical sides get smaller; eventually they become the face with diagonal BC. The largest slice then goes through the "poles" E and F of figure 8-27. All slices have the same angles (see figure 8-30) and so now we need to find the lengths of the sides.

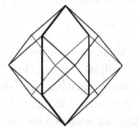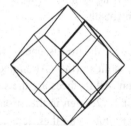

figure 8-28

The length of the sides follows by drawing two rhombi thus, with A, B and E corresponding to the vertices in figure 8-27:

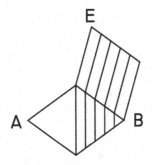

figure 8-29

The complete set of slices is shown in figure 8-30. The angle at the top and bottom is the angle of the rhombus. You can see that the slice on the right is almost a rhombus.

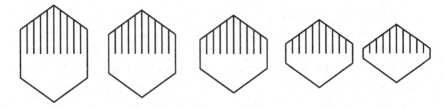

figure 8-30

The complete model is shown in figure 8-31. Because of the symmetry, when it folds flat it the shape is identical in both directions as a rhombus identical with the one from the rhombic dodecahedron, but with the tips cut off.

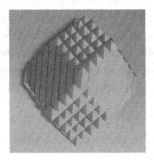
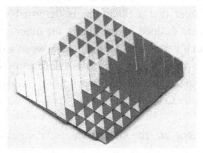

figure 8-31

Parallelopipeds

If you make a cube using drinking straws for the edges, joining them by threading string through the straws, the cube will not be rigid. It will be deformable, but not in a regular way. If it were possible to deform it so that all faces remained parallelograms then the parallelopiped created can also be used as the basis of a number of Sliceforms.

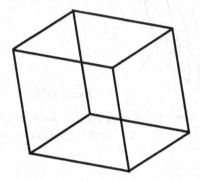

figure 8-32

The construction of parallelopipeds from rhombi can be looked at in another way. The rhombi have two angles and three angles meet at a vertex. So different combinations can occur. Denoting the angle sizes by B (big) and S (small), it is possible to have vertices with angle combinations of SSS, BBB, BSS and SBB.

Defining one vertex as BBB or SSS excludes the adjacent vertex from having the same combination, since in a rhombus B is always next to S, and also makes the diagonally opposite vertex have the same combination. Thus there are two parallelopipeds that can be made from the same shaped rhombi.

But there is only one Sliceform model that can be made. Why? Well, remember that a Sliceform is deformable and whereas two sets of parallel sides are defined by the slices, the other one set is not. This can be deformed so that a vertex SSS becomes BSS when a SBB becomes a BBB. The rhombus on the parallel set which is not defined by the slices, but by the grid, changes its shape to match the two types of vertex. The photograph in figure 8-35 shows these two situations.

All slices are the same shape, a rhombus. So only one piece needs to be designed.

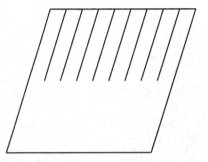

figure 8-33

The model when assembled is shown in figure 8-34 in two stances. The first where there is a pair of vertices with angles SSS and then one where the model has been deformed to be BBB. The latter is closer to being flattened and has a centre of gravity close to, if not outside the base. So it tends to fall over and flatten itself quite easily.

figure 8-34

Despite the fact that the model has the same number of vertices and faces as the cube, the breaking of symmetry from that of the cube means that it flattens in two distinctly different ways.

figure 8-35

figure 8-36

Figure 8-36 looks like a cube drawn in perspective.

Chapter 9
Variations and explorations

This chapter describes how to be creative with different methods for making models. It is a mixture of extension of ideas discussed in other chapters together with suggestions on how to create new surfaces. These may be artistic explorations, but there are also ideas which may offer new mathematical insights. Although there are descriptions of different models, the chapter is also intended to spark creative ideas.

Artistic inspiration and lateral thinking

Most of the Sliceforms described elsewhere in the book are derived from a mathematical starting point. This usually means the surface is the starting point. This might be an equation or it might be a surface to slice in order to model it. Some Sliceforms, such as surfaces of revolutions might be ones that have been created from new, in that the shape of the final surface might not be known until the model is constructed.

This chapter takes the creative aspect of Sliceforms a stage further. In some cases the shapes of the slices are predefined; in others variations on a model are made or the shape of slices in one model may suggest ways to adapt another model or suggest new ways to make models, or they may be new surfaces arrived at by serendipity, even by making mistakes. The surfaces may or may not be mathematically interesting but, like all Sliceforms, they have an aesthetic appeal. The chapter also shows how geometry can be used in conjunction with the techniques for creating surfaces as Sliceform models to make models of previously unknown surfaces.

Surfaces from circles

As described in the Preface, Henrici and Brill worked with sections of ellipsoids, paraboloids and hyperboloids which always gave circular slices. As with the other methods described in this book this relies on starting with the surface and then creating the shape of the slices. You can, however, decide from the outset that the slices will all be circles. There are then many possibilities for creating new surfaces which are easier to construct than other models. Drawing the slices is simply a matter of knowing the radius of the circles required and how the slots are placed on them.

In chapter 2, the quadric surfaces are created by using a quartic curve (a circle, ellipse, hyperbola or parabola) together with a grid. The length of the grid lines within the quartic define the diameter of a circle.

Using other curves

This method of construction opens up many ways of defining new surfaces. There is no constraint of deciding the shape of slice, only how big a circle and where the slots go. The following are start points for exploration. Although the curves do not have to be symmetrical on the grid, or have an axis of symmetry, it helps if they are because you only have to draw one set of circles, rather than a special one for every slice. Even if you use a freehand curve, mirror it to produce the symmetry.

A fish from a sine wave

Figure 9-1 shows how to arrange a pair of sine waves (they are mirror images about a central axis) together with a non-square grid to make a Sliceform model which looks very like a fish. Since the sine wave could go on forever, the end slices have fewer slots and are not part of the grid. This adds to the effect, making the end on the left appear more like a mouth and the right like a tail.

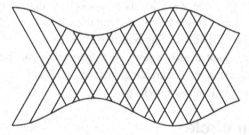

figure 9-1

It is not always easy to make the ends of the slices so that they do not fall on grid intersections. When you draw the circles for the slices, the diameters are the lengths of the lines. Use the copy method on the edge of a piece of paper (as shown in figure 3-4) to find the positions of the slits.

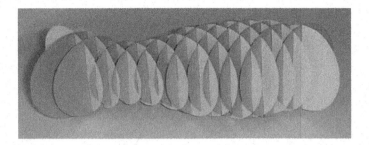

figure 9-2

The Sliceform model is shown in figure 9-2. Chapter 2 describes how such models have an elliptical shape in perpendicular cross section, and are not surfaces of revolution. This effect enhances the fish shape. A surface of revolution could be made using this shape, but it would not look like a natural fish.

A torus from circles

The following Sliceform model is different from all the others in the book. It is collapsible until the last pieces are slotted into place. Then it becomes almost rigid. Also, all slices are the same, with the slots distributed unevenly.

A torus is normally defined as a surface of revolution of a circle rotating on an axis outside it, but in the same plane. This Sliceform torus is a surface of revolution, but with the circle inclined at an angle to the plane defined by the axis and the centre of the circle. The grid is defined by a rotation as shown in figure 9-3.

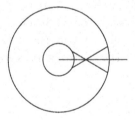

figure 9- 3

First draw the inner and outer concentric circles. Next draw a line to cut the circles at an angle to a radial line of the circles and then reflect the line about the radial line as axis. Then rotate the pair of lines about the centre of the two circles to give a set of pairs of line which all intersect in the same way as shown in figure 9-4. This means that there is only one type of circle to design for the slices whose shape is shown in figure 9-5.

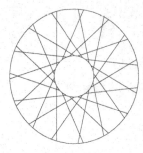
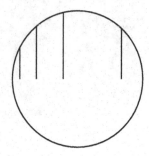

figure 9-4   figure 9-5

Because the torus is not generated in the standard way with the circle being rotated in the same plane as the axis, it does not appear to have been mentioned in the literature. Figure 4-6 shows a standard torus that has been sliced to give the set of curves known as the ovals of Cassini.

figure 9-6

Modifying other models

Once you have mastered the aspects of creating Sliceform models, then there are many other possible directions you could investigate to create new models from ones you have already made. The following ideas are some pointers for creativity.

Addition and subtraction

When you have the design of slices for a model, it is often easy to modify it by extending it or combining it with another set of slices. The most obvious case is to add a plinth to a model which has a flat base. This is equivalent o adding a part of a cube. A simple example of this is to cut a sphere in half and add a plinth so that it appears like a building with a cupola. Figure 9-7 shows the two central slices in each direction.

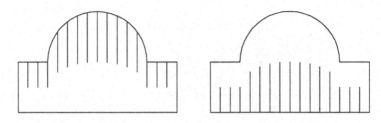

figure 9-7

The added plinth is the same as the height of the hemisphere , that is the largest (central semicircle) and that the slots have had to be drawn specially. Figure 9-8 shows the completed Sliceform.

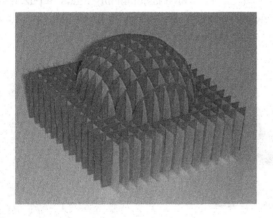

figure 9-8

It is just as easy to make a bowl by subtracting the hemisphere from a cuboid. The central slices are shown in figure 9-9 and the model is shown in figure 9-10.

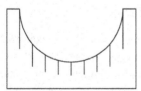 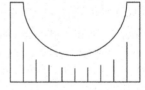

figure 9-9

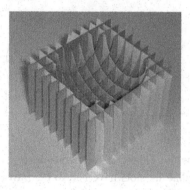

figure 9-10

A more sophisticated way of using a sphere is to cut away half of it using the method known as the *coup du roi* which is commonly found in puzzle books at the end of the nineteenth century. The puzzle is to cut an apple into two similar shaped parts by using three cuts of a knife.

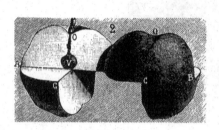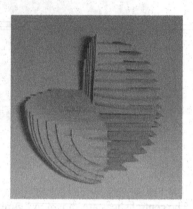

figure 9-11

Mixing surfaces

A number of models are of surfaces which are not closed, for example the functions plotted in chapter 7. The slices are there to support the surface and so the base of the model is flat. Interesting Sliceform models can be made by using the same surface underneath the plotted one. There are many variations, depending on the symmetry of the original surface:

- reflect the surface about the base
- consider the slice as an extrusion as with the profile models described in chapter 3

- rotate the surface by 90°, 180° or 270° (if its symmetry allows a difference) for either of these two possibilities
- use another matching surface on the base.

When you have designed a set of slices, converting them into a new model is often a good exercise in three dimensional visualisation. The results enhance the original model since they bring out new relationships. Figure 9-12a shows the how the profile of a sine wave (the model shown in figure 3-24) which has been adapted by adding the same profile rotated through 90°. The slices are the same in each direction. Figure 9-12b shows the conoid ruled surface shown in figure 6-21 which has been reflected about the base.

 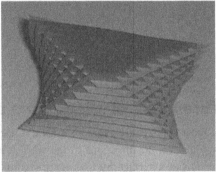

figures 9-12a and 9-12b

The symmetry of the reflected conoid surface makes for an attractive model and one which looks different from the original surface because the way it was constructed is much harder to analyse.

Cutting away parts of slices

Most Sliceform models are simply complete slices with the shape of the slice defined by its edge. Some models can be enhanced by cutting away part of this slice. There are a number of ways to do this. The hyperbolic paraboloid model shown in figure 6-14 has been produced by translating the top edge of the slice to define a strip; this is limited to models where only the top surface is being modelled.

Another method is to take a knife or scissors to a completed model. When you have taken the time and effort to create a model, you would probably think that the last thing you would want to do is "destroy" it. Where you have a simple model like a sphere or cube, which is easy to make, it can be a useful way to find new ideas. You may find that you arrive at a model you have created another way. Figure 9-13 shows a cube that has been flattened in its widest version and an ellipse drawn to fit the rectangle. Cutting round the outline, the result is part of a cylinder like the one designed in chapter 2 (figures 2-22 and 2-23), but with most ellipses partly cut away.

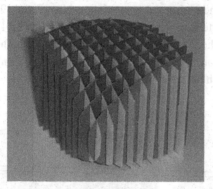

figure 9-13

Three sets of slices from two

In most constructions there is a natural way to make the pairs of slices and this way defines the directions of the slots. It is often easy to make more than one version of a surface by looking at slicing in directions parallel to each of the coordinate axes, instead of the two chosen. Many models are symmetrical in that the two sets of slices defined by the construction are the same shape, but in general there are three different sets of mutually perpendicular slices possible. Taking them two at a time gives three models. Once you have two sets of slices, you can change the direction of the slots, and then use the method for finding the slices in the other direction by using the method described in the section "Where you have slices only in one direction" in chapter 1. The construction method often defines the way the shapes of the slices are generated; for example a surface of a rotation naturally gives slices parallel to the axis of rotation.

In the case of surfaces of revolution, the shapes of the surfaces are defined by the nature of the construction. However, in all cases, if you slice them in the direction perpendicular to the axis of rotation, you will get a set of circles. Having created a particular Sliceform that is a surface of revolution, it is easy to create an extra set of circular slices and re-mark the slots in the slice shapes from the ones from the surface of revolution.

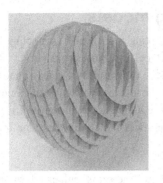
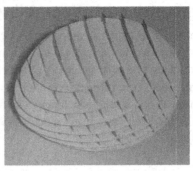

figure 9-14   figure 9-15

Figure 9-14 shows a photo of an egg shaped Sliceform that has been made as a surface of revolution of a Cassinian oval shown in figure 7-14. The resultant model created by combining the shapes of a set of original slices with circular sections is shown in figure 9-15. The visual appearance of the model is quite different when the surface is made in this way, even though the surface being modelled is the same. The way it folds flat is, of course, quite different, and the fact that the slices are not the same shape make it interact with the light in a way which leads you to question that they are the same surface.

Figure 9-16 shows two slices of this model. Compare these slices with the two slices created for making the model where the slices have slots parallel to the axis of revolution which is shown in figure 9-18.

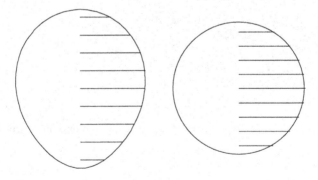

figure 9-16

Symmetry and assembling models

Symmetry is an important aspect both of creating models and assembling them. The section later in this chapter on "A surface from an astroid curve" shows how thinking about the symmetry of a surface can simplify the way a model is designed and how different versions can be made by using a symmetrical approach as opposed to just slicing a surface along two coordinate directions. Even in the latter case, many models have slices which are the same in both directions.

Most models are more attractive if the symmetry is enhanced by using different coloured card in each direction. This gives different appearances as you turn them around and helps to emphasise the play of light and shade as they are rotated. It also shows different patterns when they are flattened. These effects can be explored further in some models. Where models have the same shaped slices in both directions you might like to try swapping these pieces to change the colour effects. A more interesting aspect of symmetry occurs when the colour of the slices of a model adds chirality (that is it can exist in right and left handed forms which are mirror images of one another). One such model is the tetrahedron.

Figure 9-17a shows two tetrahedrons (see chapter 8 "Polyhedra") made from six slices which have been made with slices which are coloured different in each direction. Arranging the slices in a different way yields the mirror pair.

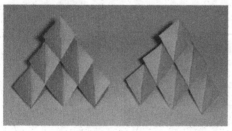

figure 9-17a

When the tetrahedra are flattened as in figure 17b, the colours are reversed.

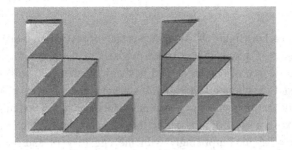

figure 9-17b

Creating new surfaces

The following method is an illustration of how serendipity can give rise to totally different and new surfaces which could not have been thought of without constructing a Sliceform model. It arose simply by making a mistake in assembling a surface of revolution. When you make a surface of revolution, by the nature of its rotational symmetry slices in each directions have the same shape. To make the model, you make the slots in opposite directions as shown by the example in figure 9-18. This shows a slice from the same position in each direction. This means that when the model is assembled both shapes face in the same direction in order that the model has rotational symmetry. Note that these slices have a vertical axis of symmetry, but *not* a horizontal one. The slots have been constructed so that they are half way up the slice at each position.

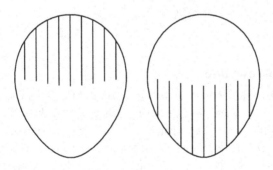

figure 9-18

Now for the variation. It should be pointed out that this does not work for all surfaces and the mathematical conditions as to when it will occur have not been worked out, but as the examples below show, this does not preclude making models which have different artistic effects.

A reverse egg

An egg is described earlier (figure 9-14) as the surface of revolution which is created from an oval of Cassini. If you make two sets of the slices for one direction of this egg surface of revolution and fit them together, you will find that they appear to slot together perfectly, but this is not a mathematical proof. You would expect the central slots to fit, since their slots match, but you might expect that as you move away from the centre then the correspondence of the slots would no longer align. Surprisingly with the egg they fit and a new surface is born. In this case, the orientation is rotated by 180° as you rotate the surface through 90°about the vertical axis of the slices. The complete Sliceform model is shown in figure 9-19.

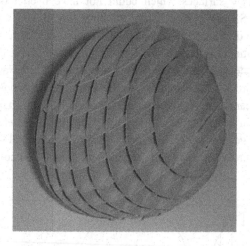

figure 9-19

The paraboloid of revolution

In the case of the paraboloid of revolution (see chapter 5), it is easy to prove that the surfaces will fit when they are reversed because all the slices of a paraboloid of revolution are the same parabola. So the surface resulting from reversing the slices (figure 9-20) is the one that results if two parabolic cylinders are intersected so that the common volume is the new surface. A parabolic cylinder is created when a parabola is translated along an axis perpendicular to the plane of the parabola. This is analogous to the way a circular cylinder is made by moving a circle.

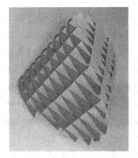

figure 9-20

Modelling vaulted ceilings

Surfaces made from intersecting cylinders have a long architectural history. They have their origins as far back Roman circular arches and are effectively the combination of two circular arches at right angles. Thus they might commonly be found at the junction of two corridors, although, because they form a very strong ceiling, they can be used anywhere. More complex vaulted ceilings, especially when fans are added, form the high point of gothic architecture.

If you take two cylinders and intersect them as in figure 9-21 and then just take the part which is common to the two cylinders, the result is shown in figure 9-22. This surface is a complete surface from the intersection. A vaulted ceiling is formed from the top half of the surface of figure 9-22 cut along the longest set of lines.

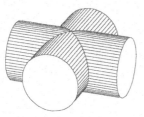

figure 9-21   figure 9-22

There are a number of possible ways to slice the surface of figure 9-22 to make a Sliceform model. The way the lines are drawn in the figure implies one of a set of slices which lie parallel to a horizontal plane (the plane defined by the axes of the cylinders). These slices are all squares and the other set of slices would be vertical slices. For construction purposes I will call this model 1. These vertical slices could also be created in two directions and the lines in the figure could then be seen as top edges of these slices to give model 2. For model 3, another way to slice is parallel to the intersection lines of the cylinders. This is the most economical in design terms since the slices are the same in both directions. The three models, as you would expect, look and behave differently, especially the way they fold flat.

For models 1 and 2, the shapes of curved slices are arcs of circles. I leave these as an exploration exercise for the reader. The following description of the geometrical method of construction for model 3 relies on drawing ellipses. The slices are part of ellipses for the same reason that elliptical sections are necessary for creating a circular cylinder as described in the section "Making a circular cylinder" in chapter 2. For that reason, the ellipse is the same for each slice; just different parts of it are used. The ellipse is defined by the intersection of the two cylinders, so the central slice is that ellipse (figure 9-24 left) and its axes are in the ratio of 1 to $\sqrt{2}$.

Figure 9-23 shows the construction diagram with four halves of the six slices. The complete slices are shown in figure 9-24. The construction begins with the central square. The grid on this square defines the size of the slices. The first half slice shows half of the central ellipse. For the other slices, the length of the grid lines define how much of the ellipse is used in each slice. Measurement takes place from the end of the major axis. The position of the slots is determined by the grid intersection points.

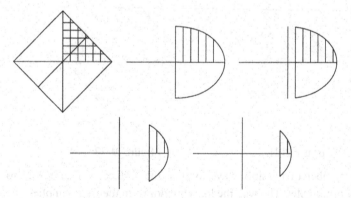

figure 9-23

Once the half slices are constructed, they are reflected to give the complete slice. The set of different slices are shown in figure 9-24.

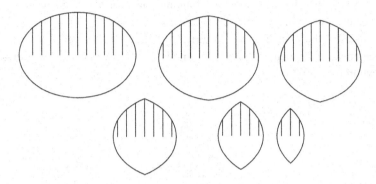

figure 9-24

These slices make the complete surface as shown in figure 9-22. To make the normal vault, then these slices need to be cut in half along the major axis. The slots then need to be halved and two sets of slices made with one having slots up and one down. Because of the symmetry of the complete surface of figure 9-22, then the slices are the same in both directions if you make the model from the complete slices of figure 9-24.

Figure 9-25 shows the different ways of constructing the Roman vaulted ceiling. The properties of Sliceforms, and the way different directions of slicing yield models which look and behave differently, mean that they need to be studied carefully for it to be obvious they are all based on figure 9-22.

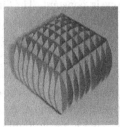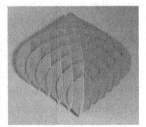

figure 9- 25

A surface from an astroid curve

This example shows how creative thinking can turn a way to draw a two dimensional curve into a three dimensional surface and is another way of how a surface can be modelled in two different ways. It is derived from a curve called an astroid which has four cusps and is commonly drawn in two ways as shown in figure 9-26.

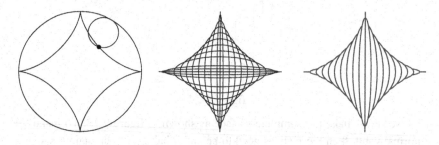

figure 9-26

The curve belongs to a family of curves known as cycloids which are formed when a circle rolls on another curve. The name cycloid belongs to the curve described in chapter 3 which is obtained when a circle rolls along a line and is described there in creating a profile surface. When circles roll on circles, the curves are known as epicycloids (when the circle rolls outside another circle) or hypocycloids (when it rolls inside). The astroid is a hypocycloid, the locus of a point on a circle which rolls inside another circle when the circle is a quarter of the size of the one in which it rolls. This is shown at the left of figure 9-26. It is also the envelope of a family of ellipses, centred on the same point, the sum of whose axes is constant. The limits of these ellipses are a horizontal and vertical line. The ellipse construction is shown in the right two parts of figure 9-26 and this method is the basis for constructing the surface.

Consider the ellipses of the centre diagram in figure 9-26. If you imagine that the ellipses are not in the plane forming the curve, but in space so that the vertical line is closest to you and as you move through the ellipses they move away from you the horizontal line is farthest away. Then each of the ellipses can be seen as slices of a surface, and would appear like the diagram at the right of figure 9-26. These concepts suggest two ways to make the surface:

- by drawing a set of slots on each ellipse of the diagram of figure 9-26 and then using the method for finding the slices in the other direction described in the section "Where you have slices only in one direction" in chapter 1.
- by modifying the slices of the tetrahedron model described in chapter 8, "Polyhedra".

The first method needs a further constraint, since the definition above does not define the spacing of the elliptical slices. Taking the tetrahedron shape as the constraint, then it is easy to calculate the total distance from the vertical to the horizontal line and define the spacing by the number of ellipses. Figure 9-27 shows two elliptical slices from each end of the model, with the central circular slice and three slices in the other direction showing how the central vertical slice is a triangle and the others are interesting tear-drop shaped curves.

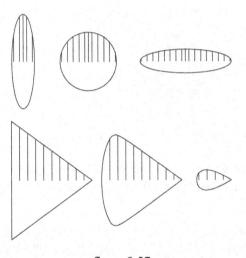

figure 9-27

Figure 9-28 shows how to modify the slices of the tetrahedron to make this astroid surface. The rectangles are used as the bounding box to define the ellipses with the central square defining the central circle. Because it uses the symmetry of the tetrahedron, the slices are the same shape in both directions.

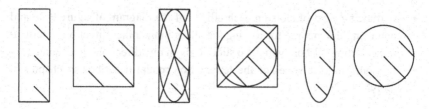

figure 9-28

Note that the slots are cut in the same way as in the tetrahedron, which means that they are not parallel to the axes. These slices are for the simple tetrahedron shown in figure 8-7.

Figure 29 shows the astroid created using the slices in figure 9-28.

figure 9-29

Figures 9-30 and 9-31 show the astroid created by modifying the tetrahedral slices, with 14 slices in figure 9-30 and six in figure 9-31.

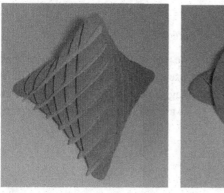 

figure 9-30   figure 9-31

Contrast the flattened version of the six slice astroid surface in figure 9-32 with the flattened tetrahedron, figure 9-17b earlier in this chapter.

figure 9-32

Inversion in a sphere

Sliceforms offer a simple method to study the geometrical transformation known as inversion. Inversion is a valuable tool for solving problems with a simple construction which are often very difficult otherwise. See Ogilvy for this and an approachable discussion of inversion. Inversion in a circle is a rather strange type of 'reflection' where all the outside of the circle finishes up inside and what was the inside now covers all the outside. Inversion is an appropriate word because inverting the inverse gets you back to where you started. Any point on the inverting circle (sphere) (the circle (sphere) of inversion stays where it is under this transformation. Very little has been written about inversion in three dimensions, but the methods for plane geometry can easily be adapted.

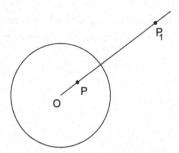

figure 9-33

In a plane, inversion is a transformation where a point P is mapped with respect to a circle such that (figure 9-33) $OP.OP_1 = r^2$ where r is the radius of the circle. The most important properties of inversion are:

• straight lines not through the centre of inversion invert to circles through the centre;
• straight lines through the centre of inversion are self inverse;
• circles not through the centre invert to circles
• circles which pass though the centre invert to straight lines.

In space, the transformation is achieved by using a sphere. A similar set of properties hold:

• planes not through the centre of inversion invert to spheres through the centre;
• planes through the centre of inversion are self inverse;
• spheres not through the centre invert to spheres;
• spheres which pass though the centre invert to planes.

The most important notable point, used in the following examples of inversion to create Sliceform models, are that circles invert into circles in space also. It is important to realise that this means a circle as a curve. A circular disk inverts into the cap of a sphere with the boundary of the disk inverting into a circle. In fact all the plane properties in the list above are also valid.

The methods in chapter 2 and the section "Surfaces from circles" at the beginning of this chapter offer a simple method to invert a number of surfaces since, if the slices of the original Sliceform are circles, then so are the slices of the model after inversion. The shape of the inverted surface depends on the orientation of the original surface to the inverting sphere. Because of the high symmetry of the sphere and the ability to use circles to create the model, then the calculations for the transformation can take place in the plane.

Figure 9-34 shows the transformation of an ellipse in two ways which is used to create two different Sliceforms as described below step by step. Note that the shape of the curves are totally different. In the final model, they represent the shape of the cross section of the model. This diagram also shows how the grids are inverted and that they are no longer sets of parallel lines. This has important consequences for the way the Sliceform folds flat. It is also important to remember that the lines of the grid are not inverted (for they would become arcs of circles), only their ends. This is described below.

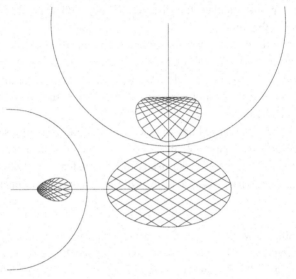

figure 9-34

In terms of creating a model, figure 9-34 shows a plane through the centre of two spheres with the cross section of an ellipsoid outside these spheres. The ellipsoid is placed symmetrically with respect to the spheres, so that we are looking at the central cross section of the ellipsoid also. The lines from the centre of the ellipsoid show axes which are joined to the centres of the spheres and the ends of these lines are the centres. In efficiency of making a model, the centre of each sphere is placed on the axes of the central ellipse of the ellipsoid. This means that the inverted surface is symmetrical about an axis and consequently the number of slices that need to be designed is halved.

Although it is possible to construct an inverse point geometrically with ruler and compass, in this model it is more accurate to use calculations either with a pocket calculator or a computer spreadsheet. The following process also describes use of a computer CAD package to draw the inverted objects. The procedure is as follows:

1. Place the inverting circle with centre at the origin to simplify the calculations.
2. Draw the ellipse and add the grid. These lines are the diameters of the circles which define the ellipsoid.
3. Use the CAD package to measure the coordinates of the ends of the lines of the grid.
4. For each point calculate the distance of the point from the origin using Pythagoras theorem.
5. Calculate the inverse position of each point and draw a line between the two inverse points of the ends of grid lines. These are the ends of the diameter of the inverted circles. We are only concerned with the ends of these lines since we only need two points at the end of the diameter to define the circle.
6. This gives a grid whose lines we know are the diameters of a set of circles. As described in chapter 2, these can be used to create the slices. The spacing of the grid lines also needs to be transferred to the circles since they are not regular. Because of the symmetry axes, not all lines need to be inverted.

It is advisable to work with a print of the grid when assembling the model, since the irregularity of the grid means you often need to check the orientation of a slice. Figure 9-35 shows an enlarged version of the egg shaped inverted ellipse and a slice to show the irregularity of the slots.

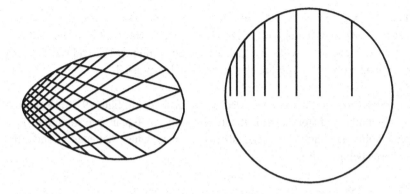

figure 9-35

As figure 9-36 shows, this Sliceform will flatten in one direction, but has restricted in movement in the other direction.

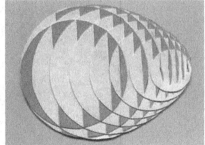

figure 9-36

The reason it will only go flat in one direction is because the grid is irregular. Figure 9-37 shows a cell forming part of the grid

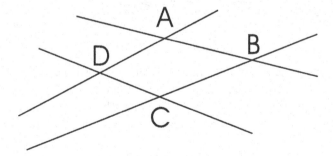

figure 9-37

This cell will fold so that A comes down to C, but D cannot go to B. This is because AD is 65 units, DC is 58, CB is 88 and BA is 80. So the flattened length AD combined with BA is close to the combined DC and CB, but DA combined with DC is very much smaller than BA plus CB.

The Sliceform for the other inverted ellipse is shown in figure 9-38. Because it has a similarly distorted grid, it too will only go flat in one direction. It is remarkably similar to the egg shaped surface when flattened despite the surfaces being so different.

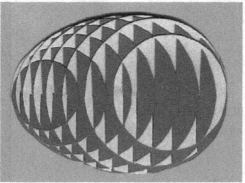

figure 9-38

Another interesting inversion is the elliptical cylinder. To get the complete surface, it is necessary to have an infinite number of circles, but even a few show the surface and the way it curves round in a ring. This surface is similar to a Dupin cyclide which is formed by inverting a torus (see Fischer page 28 for more details).

The grid for inverting the elliptic cylinder is the one shown in figure 2-20. Figure 9-39 shows how the cylinder represented by the two lines which is tangent to the circle is inverted in the circle on the left to give two circles which are tangent at the centre of the inverting circle. These two circles are the limiting circles of the section of the inverted solid which is show as a grid diagram at the right.

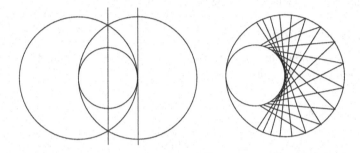

figure 9-39

Figure 9-40 shows the resultant Sliceform which is superficially similar to the inverted ellipsoid, when open, but different when flattened. It too only flattens in one direction.

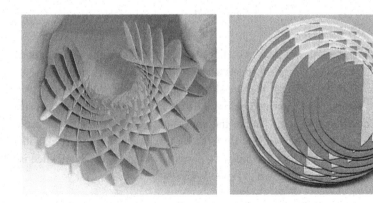

figure 9-40

The cyclide has a very strong mechanical tendency to fold flat which is why it is held open in figure 9-40.

Non parallel slices

All models, apart from the ones described in this section, rely on a set of slices and slots which are parallel. Other models are possible if the direction of the slots intersect in the same point. There still need to be two sets of slices. Whereas most other models can be considered as based on a square grid which is extruded to a cube, these models are based on a square based pyramid. This system can then be used to create other models.

The basic Sliceform

Figure 9-41 shows the steps to designing the slices for a model which is equivalent to a cube, but with non-parallel slices.

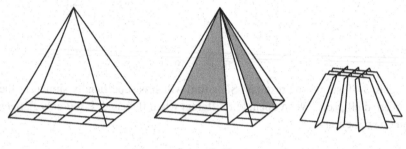

figure 9-41

The centre diagram shows how two slices would intersect if the full pyramid were used. If all the slices were added to the grid then they would all intersect at the apex. This would mean that the slices would interfere with one another. So the pyramid must be truncated before the slices are created to give the a set of slices shown at the right. Unlike the equivalent slices for the cube, each slice is different as you move away from the centre. In this simple model, with the fourfold symmetry, there are only two slices to design. The central slice is the easiest to construct as shown in figure 9-42.

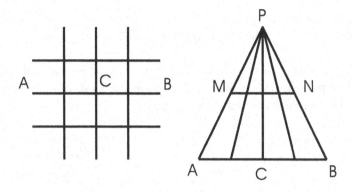

figure 9-42

The base of the slice is obtained from the grid line AB. The height of the pyramid, CP, determines the height and in this case has been chosen as equal to the width of the base. The divisions of the base come from the grid and these are joined to the apex P to give the position of the slots. The slice shape when the pyramid is converted a frustum is AMNB. The height of the slice has been chosen to be half the height of the pyramid.

A slightly different approach is required for inclined slices: see figure 9-43.

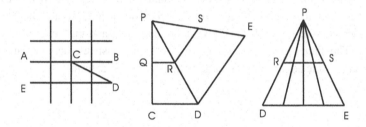

figure 9-43

The central diagram of figure 9-43 is used to construct the shape of the triangle. As with the central slice CP is the height of the pyramid. The length PD of the side of the triangle from the end of he base line DE to the apex P is found by using an auxiliary triangle PCD which is a right angled triangle with two sides known. Having constructed PD, then the shape of the triangle PED is easily found since ED is known from the grid and the triangle is isosceles. The height of the slice must also be constructed since it is slightly smaller than the height of the central slice because it is inclined to the base. The height CQ is half the height of the pyramid. The divisions of the base come from the grid and these are joined to the apex P to give the position of the slots.

Figure 9-44 shows the two types of slice with the slots marked in the normal way, with the central slices at the left, together with a pair of slices where the number of slices in the model is much greater.

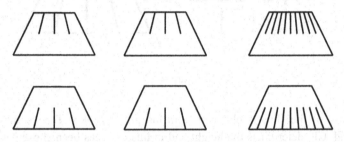

figure 9-44

This model is harder to assemble because the spacing between the slots varies. With the simple model described above, this is not that much of a problem, but as the number of slots increases this becomes a model that you should not attempt without experience. Fitting the two central slices is easy, but when you come to the outer ones, the length of the top and bottom of the slice is quite different, but the number of slots is the same. This gives rise to interesting effects when the model is flattened. Unlike other models, the base edge becomes curved. Figure 9-45 shows a simple model and then one with more slices. The flattened version of the one with larger slices, demonstrates the curvature of the edges not seen in any other type of Sliceform.

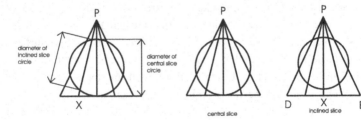

figure 9-45

A more advanced version - a sphere

In designing the slices for other models where the slices are not parallel, the same pyramid is used and the slices for the frustum of the pyramid are used to determine the slots. The following method for making a sphere explains this.

The sphere needs to have a diameter which is higher than the frustum of the basic model. The left diagram in figure 9-46 shows the central slice of figure 9-42 with the sphere cross-section imposed on it. This defines the size and position of circles on the inclined slices.

figure 9-46

The central diagram of figure 9-46 shows the central circular slice, and is essentially the right diagram of figure 9-42 together with the sphere circle. The right diagram is the right diagram of figure 9-43 with a circle whose diameter is defined as shown in the diagram on the left of figure 9-46. The position is the same distance along line XP in both diagrams. This gives slices with slots as shown in figure 9-47.

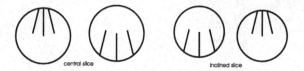

figure 9-47

The model is shown in figure 9-48. When it is flattened, the rotation effect with the flattened pyramid model of figure 9-45 is not as pronounced because of the curvature of the circles.

figure 9-48

Solid models

The beauty of most Sliceform models is in the fact that they can be flattened and the slices show a wide range of effects when they are viewed from different directions. However, you may sometimes want a solid surface and Sliceforms can be a route to this because they are an easy way to model. Solid surfaces can be modelled by filling the holes with plaster or other modelling material. The solid can be made lighter if the gaps are filled with another material like folded up paper. This also helps to keep the model from distorting as you add the plaster.

Making models rigid - a Sliceform chair

There are many possible applications for the Sliceform technique of modelling and construction. One common use was in a tray to separate fruit during transport. The fruit sat in the holes between the slices. Rigid plastic has been used to make chairs. Whereas the fruit separator are deformable and kept in place by the tray and fruit, for a chair, rigidity is required. If the plastic is thick enough then the slots cannot act as hinges. Another solution for collapsible chairs was devised by the mathematics student Rhys Williams and placed on the internet. This chair can be made rigid as the following description shows, and then folded up for storage or transportation.

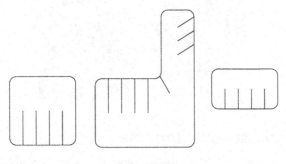

figure 9-49

The slices for a slightly modified version of his chair are shown in figure 9-49 and consist of the two types of slices at the left which form the chair (four of each required) together with four locking slices as shown at the right. If the locking slices are not inserted then the chair can fold flat. The chair is made of thick board, with sufficiently thick slots to allow folding. Note also that a cover is required on the seat because the points of contact of the edges of the slices would otherwise make it uncomfortable to sit on for long periods. The photographs in figure 9-50 show Rhys Williams' chair.

figure 9-50

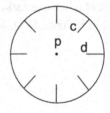

Parametric surface models

In chapter 7, the section "Parametric equations and polar coordinates" describes how some surfaces can be defined using a non-cartesian grid and how the resultant slices interact so as to make the models rigid. These models are not Sliceforms, but perhaps more Surforms because they do not fold flat, but nevertheless the design is based on slices of surfaces. The following example shows how this type of model can be created. This method is suitable for making surfaces of revolution directly and much easier than Sliceform constructions.

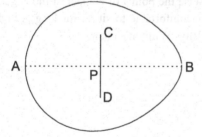

figure 9-51

These models require a centre support as at the right of figure 9-51. The shape of the curve forming the surface of revolution is reflected about line AB to form one pair of half slices through the axis. The size of the centre support and line lengths depend on the model, but will become apparent from the construction steps. In this model four of these pieces with one centre support are required and the pieces are cut and assembled thus to give the model shown in figure 9-52.

1. Cut out the centre support and cut slots as for Sliceform models along each of the marked lines, so that the slot is as thick as the card.
2. Cut out the rotation pieces and fold along line AB. Cut along line CD, but only cut a slit and not a slot.
3. Open out the folded rotation piece so that it is approximately the angle of *pcd* in the support piece. Gently open the slit CD and push it onto the centre support piece so that point P reaches the centre When constructing the line lengths CP on the rotation piece equals cp (the uncut portion) on the centre support. The radius of the support piece can be the same as the length of CP extended on the rotation piece or less as in figure 9-51 and the model in figure 9-52.
4. Adjust the position of the folded halves so that they fit along the angle and uncut parts of the rotation piece fit in the slots of the centre support. When you release the opened slit CD in the rotation piece, it acts like a pincer to grip the centre support. That is why you must only cut a slit in the rotation piece. You need slots in the centre support to take the width of card though.
5. Complete the model by attaching the other three pieces in the same way, so that each rotation piece occupies two adjacent slots of the centre support.

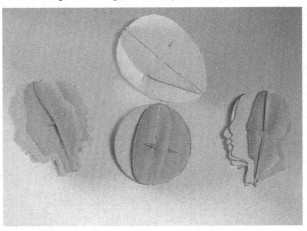

figure 9-52

The photograph of figure 9-52 shows an egg, a sphere together with another pair of models where the profile of a head has been used for the curve on the rotation piece. In the one on the left, just the face profile is used, whereas on the right, the slices are either faces or backs of the head.

Figure 9-53 shows a collection of models from the Schilling catalogue. They are a group designed by the mathematician C Tesch to show the three types of curvature of surfaces as described in chapter 4 (see figures 4-7 and 4-8). Each model shows twelve circles of curvature with curves of intersection of the plane formed by the circle of curvature with the planes which are normal to the surface.

figure 9-53

Chapter 10
Using the Computer

You can use computers in many ways to create Sliceforms even if you are not a programmer, although it obviously helps to be computer literate.

Important points for reading this chapter

The previous chapters in the book are about taking the geometrical and visualisation steps required and making the models. The chapter is not meant to be a set of practical examples of making models using software, but a survey of what type of software you might use and why. It also provides additional tips for some specific software packages which apply to similar ones.

The chapter falls into three sections:
- a general survey on types of programs available and where in the design and production process they could be used;
- some general ideas for programming;
- a description some packages which allow you to plot surfaces which are mathematical functions.

This last section might interest you if you are particularly interested in chapter 7, but most Sliceforms are *not* about plotting but visualising and drawing.

In many cases, I have had to invent methods of using the software in order to make some types of models, so the chapter is also a personal approach to using software. I was concerned with the model rather than writing a program for others to use. So the techniques are a creative (perhaps idiosyncratic) collection of methods for using different combinations of packages which allow manual intervention.

Some parts of this chapter describe programming and specialised use of software. If you not a programmer or a user of computer graphics, just ignore anything which does not apply to you. Some of the methods are specific to a particular type of package. Whereas the rest of the book is probably timeless, this chapter will date quickly but I hope many concepts will be useful in the future. Keep looking at the website (see *Where to find out more* at the end of the book) for further developments.

Using the computer to create the slices

Sliceform models do not need a computer to produce them, but there are advantages in using one. If you want to make more than one version of the same model, you can just print one out without having to redraw it. If you make a mistake, or want to make a modification, then you only have to work on the part that is different and can make a copy of the original to work on. You can also develop different versions once you have made a suitable original. If you are not able to draw sufficiently accurately by hand, then the ability of a computer package or program to draw straight lines can mean the difference between a good model and one which does not quite fit together properly. Another benefit of the computer is in instantly being able to change the size of the model, for example to fit on the card you have available. But the main reason for using a computer, rather than drawing by hand is that you correct mistakes easily and tidily and that the whole process is so much faster.

However you use the computer, like many other tools, it helps if you have been through the process using conventional methods, since you understand what is happening more intimately. In designing your own models, it also helps if you have constructed some, by using the templates at the end of the book, and understand the three-dimensional version of the object you are trying to make.

The slices for all models I make are drawn on the computer in one way or another. I very rarely use the same package from start to finish. I use a mixture of packages for different reasons. The examples in this chapter describe some of the ways I use them.

The main packages I use fall into three groups.
- For geometrical drawing, I use the two dimensional drawing program *Autosketch*. This is a simple CAD (Computer Aided Drafting) package.
- For fitting a drawing onto card for printing, adjusting the line thickness, and tidying up slices from a program I have written, I use the graphics design package *CorelDRAW*.
- For programming slices, I use the DOS based *Microsoft QBASIC* in conjunction with CorelDRAW.

For special models, as described below, and to produce diagrams in three dimensions, I would recommend the following.
- For three dimensional work, DesignCAD;

- For plotting implicit equations, program *GrafEq* from Pedagoguery Software.
- For plotting three dimensional surfaces using algebraic equations, I used a number of packages available on the Internet as shareware which are often written by college students as an exercise; two I have found particularly useful are Clark Dailey's Plot3D suite of programs (which can plot explicit, implicit and parametric programs) and David Parker's DPGraph.

See the bibliography at the end of the book for more details of how to get software from the internet.

These are just packages I happen to know or have learnt and not had time to move on, or they are simple enough and inexpensive enough to fulfil my needs. I do not use all their features. The following descriptions show how and why I need, or find it convenient to use each type. There are many more equivalent ones and, no doubt, better ones will be available in the future.

One important point to make is that "paint" packages, suitable for use with photographs, like Adobe Photoshop, are *not* suitable, except in special cases for very limited conversion between some of the other software. CAD and design packages work with lines and curves as objects, whereas paint packages work with pixels, and you need to be able to deal with lines and curves as you would if you were drawing manually.

CAD packages use the manual drawing methods described in the rest of the book, but before going on to describe other computer methods in practical detail, the following reviews of the different types of software might be helpful in understanding the possibilities of using the computer.

CAD packages - pros and cons

The benefits of using a CAD packages are that they are designed to work geometrically. This is not the case with graphics design packages, although some of them can be configured to come close. The main benefits of using two dimensional packages are:
- they are designed for working geometrically and Sliceforms are a geometric ;
- "perfect" straight lines between points and undeviating right angles
- the ability to "snap" to the end of a line, or place an object on a grid, giving an accuracy impossible by hand;
- the facility to measure the half way points of lines for making slots instantly is easy because the package finds midpoints of lines instantly

- where more than one copy of a slice is required, there is no need to draw another one;
- when an object needs to be mirrored about a line, this can be achieved exactly; design packages do not do this with their mirror option, they simply flip an object, that is turn it over on top of itself;
- the situation described in chapter 1 in the section "Where you have designed slices in just one direction" when only one set of slices can be constructed geometrically is easier to work with since parts of a set of slices you have constructed can be made quickly by copying and placing in a regular fashion accurately;
- the packages have the ability to export and import different file types for using the output with other programs.

The main disadvantages are that they frequently do not have the ability to change line thickness and a good preview for printing, particularly in a draft and final production mode. These and other disadvantages are outweighed by the benefits of the design package.

I use three dimensional CAD programs more for drawing and displaying three dimensional views of objects. Some packages have the ability to slice solids, but I have not found them easy to use for making Sliceforms. The reason I use DesignCAD is that it is the only package I have found (including high end packages like AutoCAD) which will produce accurate hidden line elimination so that the intersection of the slices are shown. Many of the drawn figures of three dimensional objects in this book have been produced in this way.

Graphic design packages - pros and cons

Packages like CorelDRAW and Adobe Illustrator are an essential part of the way I produce the final output of a template. They allow objects to be manipulated as lines and circles, for example, in contrast to the other type of graphics packages which work with photographs at a pixel level.

The main benefits of using graphic design packages are:
- individual line thicknesses can be changed easily;
- they have a "draft" mode which instantly allows the variation in thickness to be reduced so that accurate joins can be made;
- they are much easier to work with when laying out a set of slices to fit on a piece of card;

- they have the ability to trace around the outline of a curve, or along a line, a feature which I have used for some algebraic surfaces as described in the section "Creating a surface from an implicit equation" later in this chapter;
- for more sophisticated models, produced by programs, they are able to import files from the program I have written so that the slices can be tidied up.

This last aspect is described in the section below on programming. I program to produce slices for models rather than produce a program that can be used by anyone. There are so many different types of models that it would be difficult to write one program. Arranging the slices on a page to print economically can only be done manually.

The disadvantages are that they are not geometrically based, and CorelDRAW particularly has features which render them unsuitable for geometrically construction:

- circles are not drawn using a centre and radius;
- the ability to "snap" to the end of a line, or place an object on a grid, is not as sophisticated as with a CAD program;
- they do not have the ability to snap to mid points of lines, which is vital for drawing slots;
- when an object needs to be mirrored, they do not perform a true geometrical reflection about a line, but simply flip an object in a vertical or horizontal direction.

Some of these problems can be overcome, by careful reconfiguring of how objects are placed, or altering grid spacings, but this is extremely time consuming in a situation where such settings need to be constantly changed.

Programming - pros and cons

Where the model is defined by equations or calculations are needed, then a programming language must be used. Software to display surfaces generated by equations is discussed later in the chapter, but they cannot generate the slices in a suitable form for making the models. Programming means writing in a formal language, like BASIC or C, and in some cases using a spreadsheet. In either case more technical computer literacy is required than for the use of a CAD or design package. Most programming is concerned with creating a user interface and not the end result and is extremely time consuming. I use *Microsoft QBASIC* because it is easy to program. I still need to manipulate the result by hand and this is where the features of design or CAD software have been useful. It is also possible to calculate and plot curves with spreadsheets. I find it more challenging than programming directly, and so there is only a small amount on using spreadsheets on page 235 and 242.

A design package is ideal in this respect for tidying up, for example removing "construction" lines or for making all the slices fit onto a piece of paper. There is no point in spending time on creating over-detailed programs. For example, measuring slots may be easier in a graphics program or by hand in a fraction the time it takes to write the program.

Even so, I use programming methods in specific cases, the obvious cases being where the shapes of the slices are based on mathematical equations.

Some programming examples

The programming described below is designed to produce results quickly. Where variables need to be changed, for example to create a new file or size for the slices, then I edit a hard coded value and rerun the program. By storing values in a data statement, this has the additional benefit that I have a stored copy of the values I used, not ephemeral ones that I typed in. I also keep a notebook of how and why the program works.

In the discussion below, I have described mainly techniques and algorithms, since these are independent of a programming language. The most important of these is to create files which can be imported into a design package.

Where examples of programs are shown, they are given in the Microsoft BASIC programming language which can easily be converted into others. Some of these techniques also apply to data calculated using a pocket calculator as will become apparent.

Use of program screen output

It might seem that you could just plot on the screen and then print the screen to get the slices. You might also find it easy to copy the screen to a graphics design package and then print from there. However, both these methods have pitfalls, presenting challenges such as the following.
* How do you capture the screen and then put all slices into a design package in a suitable format? Often the screen plotting and/or the capturing process give you pixel based images and not curve and line objects.
* If you have managed to capture the screen, how do you cope with having a pixel based bit-image and not a vector one?

- How do you add to or edit captured raster output?
- How do you ensure that scaling is maintained both on the screen and when slices are put in the design package? Allied with this is that printing from the screen may not give you the correct size for the slice in order to make the model. It can also be wasteful when photocopy card is used directly.

These challenges are not insurmountable. They mainly arise because of the need to have output as line objects and not pixel based ones. They are the same ones as present themselves when working with implicit equations, where there is no alternative but to plot pixel based output and are dealt with in the section "Creating a surface from an implicit equation" later in this chapter.

My philosophy has been to use different features in a mixture of packages. Writing output to a printer is a programming overhead I could not afford, simply because of the need to arrange and scale the slices efficiently on the page. So I have devised the method below for taking output from either a written program or a package like a spreadsheet into the design package which then allows me to use manual manipulation as with hand drawings.

Using metafiles to copy

You might find that you can copy the output from a plotting package (including a spreadsheet graph) in a form called a metafile which preserves the plot as an object. Look in the help for the program for ways of copying and pasting. If possible, always use the *Paste Special* option on the Edit menu in Microsoft Windows programs to paste the result into a graphics design program.

Output from programs into the design package

The following technique produces a file which can be imported into many graphics design packages and even into some word-processors (like Microsoft Word) in which you can edit the objects. (However, you must have the graphics filter to convert the file into the word processor's graphics format. This is one of the options during install and can be added later. See the installation instructions for your word processor.) The file is a simple text file and very easy to produce. It could even be used to transfer coordinates calculated using a pocket calculator into a text editor and the resulting file imported directly.

The method described below creates a file of plotter commands. The plotter commands are for the HPGL plotter language. HPGL stands for Hewlett Packard Graphics Language. Only a few basic commands are used.

- IN initialises the printer;
- SP1 chooses pen 1, or SP2 pen 2 and so on, which is useful for seeing different lines in different colours;
- PU lifts the Pen Up and moves to the point with the subsequent x,y coordinate;
- PD puts the Pen Down and draws a line from the previous plotted point to the point with the subsequent x,y coordinate;

and you also need to be aware that a unit is approximately one thousandth of an inch.

In addition, coordinates are separated by spaces or commas and a command terminated by a semicolon. Although commands do not have to be on a separate line, it often helps if you need to look at the output from your program and check it. Coordinates may be written as integers or real numbers and not in scientific notation, so you need to force the format of the coordinates. In BASIC this means using the PRINT USING command. The import filters in most design packages will accept negative numbers, although this means that the position of the imported objects will need to be moved manually to fit on the page. They usually need to be scaled as well.

A simple set of instructions to draw a line from point 100,100 to 540,670 to 800,265 and then another, different, line from 120,400 to 800,500 would look like this:

```
IN;SP1;
PU 100,100;
PD 540,670;
PD 800,265;
PU 120,400;
PD 800,500;
```

You might like to try typing this into a text file, saving it and importing it into a design package like CorelDRAW. Use the Windows *Notepad* program. You can also use a wordprocessor to create the file, but if you do, you must ensure that you save the file as a plain text file with no formatting. Save the file with the extension .PLT, which is how CorelDRAW and many other packages recognise it as an HPGL file. The following steps show how to do this in CorelDRAW; other packages will have similar functions.

1. On the *File* menu, select the *Import* option.

2. In the Import dialog box, select the directory where the file is stored and choose the file type of *HPGL Plotter file (PLT)* from the drop down menu.
3. When you have selected your file, accept the default values at the *HPGL Options* dialog box.
4. When the screen clears, the lines appear with a shape looking like that in figure 10-1

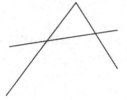

figure 10-1

You can now scale, change the line thickness, colour or manipulate the object using any of the CorelDRAW tools. If you want to manipulate individual lines, you will have to ungroup the object.

Output from a Spreadsheet

The following set of cells shows how the cells can be programmed to give the set of lines in figure 10-1. Data has just been entered in the cells for this example, but the real use of spreadsheets is to calculate the shapes of the slices. If you were using the spreadsheet to calculate coordinates for a curve, you would write the intended output in a similar format.

	A	B	C	D	E
1	IN;	SP1;			
2	PU	100	,	100	;
3	PD	540	,	670	;
4	PD	800	,	265	;
5	PU	120	,	400	;
6	PD	800	,	500	;
7					

figure 10-2

To create the file for importing into your design package as a plotter file, in this simple case it is easiest to save it as a text file. In Microsoft Excel, this means saving as *Formatted text (space delimited)*. Where you have a spreadsheet that has comments and other values which you do not want in the HPGL file, copy the cells you require to the clipboard and paste them into a text editor (*Notepad* or *Wordpad*). Then save the file as a text file with the extension .PLT.

Once you have the text file, importing into CorelDRAW is the same as on the previous page.

Output from a BASIC program

In the example program shown later, a file is opened and the printer initialised in the subroutine *PrintInit:*. Output is generated by a subroutine *xyprint:*. These subroutines are called using current values of variables and are as follows:

```
PrintInit:
    OPEN file$ FOR OUTPUT AS #1
    PRINT #1, "IN;SP1;"
RETURN


xyprint:
    IF flag = 0 THEN PSET (xp, yp) ELSE LINE -(xp, yp)
    IF flag = 0 THEN plot$ = "PU " ELSE plot$ = "PD "
    PRINT #1, plot$;
    PRINT #1, USING " #####.## "; xp * pfact;
    PRINT #1, ","
    PRINT #1, USING " #####.## "; yp * pfact;
    PRINT #1, ";"
RETURN
```

The *xyprint:* subroutine also plots the lines on the screen. The coordinates *xp* and *yp* are the values passed to the subroutine. A practical point to bear in mind is that if the coordinates are too large for the formatting definition for the PRINT USING command, then a percent sign will be output in front of the value. This may prevent the file being loaded properly. Use a text editor to search for the percent sign in the file. If it occurs, insert another hash (#) in the PRINT USING command to allow for the larger value. The constant *pfact* allows scaling the size for printing. The conditional statements use a flag to decide if the start of a line is to be plotted or if the line should be drawn, either by plotting a point to the screen (in the first case) or by setting the command to move the pen to the start of the line (PU) or to draw (PD). See the program example to produce a set of slices.

An example of using programming

The following example of plotting an equation compares programming using BASIC with programming using a spreadsheet. The logic is the same in both cases. A working knowledge of using spreadsheets and programming is assumed. Both methods can be adapted to other equations.

The example plots slices for the surface with the following equation:

$$z = x^2 - y^2$$

Analysis before plotting

The first stage in deciding how to program the slices is to understand so that the

- what the model might look like, by using a three dimensional graphics plotting program or simply by examining the equation;
- what symmetry it has, so that only necessary parts are programmed (which is particularly important where spreadsheets are concerned);
- what are the ranges to plot which also includes decisions on the number of slices.

As described in chapter 7, the convention I am using is that if the surface is sitting above your desk, the xy plane is the conventional cartesian plane on the desk and the z axis is perpendicular to the desk. Slices are plotted parallel to the xz and yz planes. Examining the example equation shows that it has reflective symmetry about the xz plane (the plane $y = 0$) and the yz plane (the plane $x = 0$), because x and y are each raised to the second power. Moreover, if you look at the shape of the curves on the xz and yz planes (with y set to zero in the first case and x in the second), then they are parabolae of different orientation; that is to say when y is zero the equation becomes $z = x^2$ and when x is zero it becomes $z = y^2$. This can be seen in figure 10-3 and in the slices shown in figure 10-4.

This leads to the conclusion that it is a saddle shaped surface. In fact the surface is a hyperbolic paraboloid since slices parallel to the xy plane have the equation of a hyperbola. The curve on these slices has the equation $a = x^2 - y^2$ where a is the value of z for the slice. [Note that the equation of hyperbolic paraboloid in chapter 7 was given as $z = xy$ and not as the equation used here, $z = x^2 - y^2$. Rewriting this as $z = (x - y)(x + y)$ shows that it is $z = xy$ rotated by 45 about the z axis.] The use of three dimensional graph plotting program shows this in figure 10-3. The hyperbolic paraboloid is also constructed as a ruled surface in chapter 6.

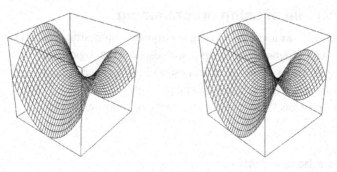

figure 10-3

Because of the symmetry, sets of slices only *need* to be plotted for x and y from 0 to +1, which gives a maximum and minimum for z of ±1. If you are programming with a language like BASIC, then it is not much effort (as in the example below) to plot the whole set of slices, but if you are using a Spreadsheet and have to manually copy a slice to another program, then being aware of such properties can save time and effort. In the example, for simplicity in showing the slices (figure 10-4), only three slices are plotted in each direction. The y-slices correspond to $y = 0$ and $y = ±0.5$, with equations z $= x^2$ and $z = x^2 - 0.25$ and the x-slices correspond to $x = 0$ and $x = ±0.5$, with equations $z = -y^2$ and $z = 0.25 - y^2$.

When you have decided on the slices that are required, you then need to program the following set of lines:
- the curve for the slice;
- the slots for the slice;
- the base for the slice;
- the vertical edges.

Other considerations are that although the range to plot is ±1, the HPGL units need to be scaled by a factor of 1000 to give a Sliceform model that will fit into a two inch side cube.

QuickBASIC version

The following program is provided as an example, following the analysis above. It plots the slices on the screen and also to an HPGL file. It is commented with additional notes following it. It has been written so that other equations can be used and magnification factors adjusted for those equations.

```
' Hyperbolic paraboloid sliceform
DEF fnSFORM (x, y) = (x * x - y * y)

file$ = "hypar.plt"
GOSUB PrintInit:

xmin = -1: xmax = 1
ymin = -1: ymax = 1
dx = xmax - xmin
dy = ymax - ymin

ncut = 4:   'size of grid
' define step for slices
stx = dx / ncut: sty = dy / ncut

m = 30:   ' number of points to plot
' define step for curve
stpx = dx / m: stpy = dy / m

fact = 100:   ' screen plotting magnification factor
pfact = 10:   ' file plotting magnification factor

SCREEN 10
' origin at centre of screen
sfact = 1:   ' size of screen factor
wx = 640 * sfact: wy = wx * 3 / 4

WINDOW (-wx, -wy)-(wx, wy)

' slice stepping across the x direction
FOR x = xmin + stx TO xmax - stx STEP stx
xoff = x * 700
flag = 0:   ' to start plotting end of line

FOR y = ymin TO ymax + .01 STEP stpy
z = fnSFORM(x, y)
    xp = y * fact + xoff: yp = z * fact
    GOSUB xyprint:
    IF flag = 0 THEN flag = 1
NEXT y

' plot baseline of this x slice
bb = ymin * fact
yp = bb + yoff
xp = bb + xoff
flag = 0
GOSUB xyprint:
xp = -bb + xoff
flag = 1
GOSUB xyprint:

' plot slot lines for this x slice
bb = ymin * fact
FOR y = ymin TO ymax STEP sty
xpp = y * fact + xoff
z = fnSFORM(x, y)
```

```
ypp = (fact * z + bb) / 2
flag = 0
xp = xpp: yp = ypp
GOSUB xyprint:
flag = 1
xp = xpp: yp = bb
GOSUB xyprint:

NEXT y
' next x slice
NEXT x

' slice stepping across the y direction
yoff = 200:  ' offset to move y slices above previous

FOR y = ymin + sty TO ymax - sty STEP sty

xoff = y * 700
flag = 0:  ' to start plotting end of line

FOR x = xmin TO xmax + .01 STEP stpx
z = fnSFORM(x, y)
    xp = x * fact + xoff: yp = z * fact + yoff
    GOSUB xyprint:
    IF flag = 0 THEN flag = 1
NEXT x

' plot baseline of this y slice
bb = xmin * fact
yp = bb + yoff
xp = bb + xoff
flag = 0
GOSUB xyprint:
xp = -bb + xoff
flag = 1
GOSUB xyprint:

' plot slots for this y slice
FOR x = xmin TO xmax STEP stx
xpp = x * fact + xoff
z = fnSFORM(x, y)
ypp = (fact * z + bb) / 2 + yoff
zp = z * fact + yoff
flag = 0
xp = xpp: yp = zp
GOSUB xyprint:
flag = 1
xp = xpp: yp = ypp
GOSUB xyprint
NEXT x

' next y slice

NEXT y

END

PrintInit:
    OPEN file$ FOR OUTPUT AS #1
    PRINT #1, "IN;SP1;"
RETURN
```

```
xyprint:
        IF flag = 0 THEN PSET (xp, yp) ELSE LINE -(xp, yp)
        IF flag = 0 THEN plot$ = "PU " ELSE plot$ = "PD "
        PRINT #1, plot$;
        PRINT #1, USING " #####.## "; xp * pfact; yp * pfact;
        PRINT #1, ";"
RETURN
```

The slices are not quite complete at the ends (see figure 10-4), but there are sufficient lines for you to cut them out and you can manually adjust the length when imported into CorelDRAW. The slices in the x direction of the equation are the ones with the slots coming from the base and the slices in the y direction are in the row above. Because the slots come from the top of the curve for the slices in the y direction, a separate base line is required when they are plotted. Other points of explanation in addition to the comments are:

1. The scaling factor for the screen is further multiplied by the plotting factor to make the range for the HPGL output appropriate.
2. When plotting the lines of the curve the range has an additional 0.1 added (for example FOR y = ymin TO ymax + .01 STEP stpy) to avoid rounding errors in the calculation of the step size and ensure that the complete curve is drawn.
3. When plotting the slices, they are offset on the screen or plotter using the variable xoff or yoff
4. The variable bb is the baseline and is set to the minimum for the range plotted but it could be moved. Slot heights are calculated using the distance of the curve from this baseline.

The resultant slices (three, in this case, in each direction) are shown in figure 10-4. They can easily be tidied up in CorelDRAW so that the lines at the edge are complete. It is usually not worth spending the extra programming time to get perfect diagrams since the effort is out of proportion to the manual effort required to make adjustments.

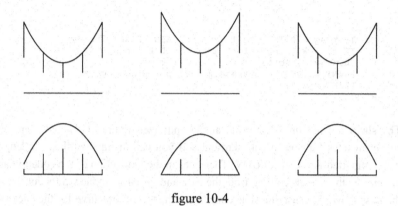

figure 10-4

Spreadsheet programming.

Once you have analysed the function, as described above, there are two ways to produce the images for the slices:
• using output from graphs
• creating output for importing into a design package directly.

Setting up a spreadsheet to do the latter can be tedious process. Simply plotting a function, or calculating slices for surfaces of revolution, is relatively simple. Adding the additional lines for slots, baseline edges and producing the output for a design package, is quite time consuming. However, once it has been done for one function, it is relatively easy to change it for others.

I have not spreadsheets to create models since programming is far easier. However, there are lessons to be learnt from the BASIC programming method. For example, do not plot every slice, but set up your worksheet to plot one slice in the x direction (keeping y constant) and one slice in the y direction (keeping x constant). To vary the position of the slices, then just alter the value of the cells which determine the constant x or y. Also, use calculations as in the BASIC program to vary the number of slices.

To use graph plotting directly, the graph can be copied and pasted into a graphics program (use *Paste Special* as described in the metafile section on page 233). Take care that all slices are the same scale. This can be achieved easily by the idea above of only plotting one x-slice and one y-slice and changing the position of the slice in the spreadsheet by manually deciding which slice to plot. You can also plot an algebraic surface in the spreadsheet.

Generating surfaces of revolution.

Programming really comes into its own when it comes to generating slices for surfaces of revolution. It simplifies programming if the program for creating the slices reads in a file of coordinates. This allows the rotating curve to be created separately; it could even be a non-mathematical one, traced from a photograph or drawn by hand and the coordinates entered into the file.

The method used is similar to the one described for drawing surfaces of revolution in chapter 5. The two sets of slices being constructed are the ones parallel to the z axis of rotation. Points of the curve are chosen in order. These points are rotated about the axis of rotation to define a circle which is then intersected with the planes of a slice in order to determine the boundary of the surface of revolution on the slice. As each point P of the curve is chosen, it defines a point Q for the chosen slice. The process must be repeated for each slice. Since the surface is a surface of revolution, Q is a point on the surface. The circle shown is the one obtained that would be obtained by slicing the surface at the position of point P in the left diagram.

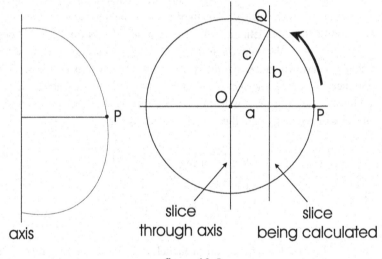

figure 10-5

The left part of figure 10-5 shows a curve from which the surface of revolution is being constructed. At the right we are determining the position of point P when rotated to point Q which is on a slice which is not through the axis. In this diagram we are looking down the axis of revolution which goes through the paper at point O. The axis is the z coordinate axis.

In the diagram on the left, we have the curve in a plane which is being viewed from the front, and we can see the curve relative to the axis. The curve starts in the xz plane. Point P of the curve and has coordinates (c, z), with c then being the radius of the circle. The z coordinate defines the z coordinate of the curve on edge of the slice at Q because the rotation does not alter its height.

The x coordinate of point P is the radius of the circle c, the circle where the curve is rotated at this height position. We are trying to find out where the curve intersects the slice when it has been rotated. We know this slice is a distance a from the axis (because its position has been chosen as the slicing plane), and so we can find the distance b which tells us where the point Q is on the slice. This is a simple use of Pythagoras' theorem:

$$b^2 = c^2 - a^2$$

If the slicing plane is farther away than the circle radius, then the circle will not intersect it. The principle behind the program is thus quite simple, but additions have to be made to cover testing to see if this intersection occurs. It also requires extra programming to calculate the length of the slots, but not to plot their position which is known from the definition of the slicing plane. This can be found quite easily manually when the output is in the graphics design program. However, in the case of a curve which has symmetry about the horizontal axis as here which leads to slices with the same symmetry, there is no need to worry about clipping the slot lines to the curve. You will cut them off anyway. They can be just plotted from the axis of symmetry until they are clear of all slices.

figure 10-6

Figure 10-6 shows a few slices from a model which is formed by rotating an ellipse. It shows how, sometimes, there are difficulties with the part of the curve at the ends of the slice. This may be because the slope of the surface is changing too rapidly for the surface to be plotted, or it may be that there are not enough points in the curve definition file to resolve that part of the slice. As figure 10-6 shows the gaps are usually slices which are farthest from the rotation axis. They are easily joined when cutting out the slice.

Another problem that can arise when the plotted slice appears jagged. All curves are plotted as a succession of straight lines. Because the curve being rotated is a discrete set of points, it often happens that there are not enough points to give a smooth edge to the slice when the curve is rotated. This can be avoided by testing if the distance between two points is above a minimum value and linearly splitting it into a succession of smaller points before plotting.

Using plotting software for algebraic surfaces

There are many sophisticated mathematical packages which allow surfaces to be plotted, such as Mathematica, MathCAD and Maple. Their disadvantage is their cost and the fact that they are general purpose mathematical packages and so often have a complex user interface. There are also some curve and surface plotting programs of varying levels of complexity available on the internet which are easy to use and either free or low cost shareware; they have the added advantage that their authors can be contacted by email should you need help. They often have minimal documentation, but since they are intended for one purpose, the user interface is simple and easy to understand and so this is not a problem. They also have a number of examples which can be starting points for surfaces to model. The following is a description of a few packages available when I was writing the book. More sophisticated ones may come along, but the principles of use are essentially the same. See the bibliography for details of where to download them.

The following are some methods I have used when creating models of algebraic surfaces, particularly implicit and parametric surfaces. As describe in chapter 7, they are also useful to decide on ranges to plot for explicit surfaces. My approach has been as follows:

- starting from an equation I want to follow, choose either Plot3D or DPGraph using various options within the programs, such as changing variables, ranges to plot to be able to visualise the model or decide that it is not suitable for plotting;
- use a program like GrafEq; to plot the equation on each slicing plane; it is possible to set the view and range in Plot3D or DPGraph so that only a slice is plotted, but I have not found the results easy to work with;
- capture or save the screen;
- trace the screen capture using a program like Corel Trace and edit to produce the slices and use CAD program for creating slots.

These steps are detailed in the various examples below. There are also some other methods using conventional programming which can simplify the mathematics, especially in the case of surfaces which are defined by a set of conditions (see the section "Programming techniques for surfaces defined by conditions" later in the chapter).

Plot3D example

The freeware Plot3D (written by Clark Dailey) is a suite of three programs that allows you to enter the equation with a simple syntax. For example the implicit function for the tear drop surface shown in figure 7-5 is:

$$0{\cdot}4y^5 + 0{\cdot}7y^4 - (x^2 + y^2) = 0$$

Various other parameters can be entered as shown in the following dialog box:

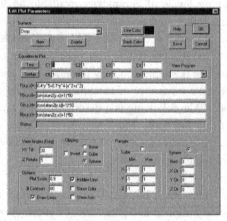

figure 10-7

There are features such as colour which are not really necessary for just examining the surface, although because the colour is plotted as a function, it can yield other information about the surface. I usually choose the option not to display colour. Much of the computer output in chapter 7 was obtained by using this program, including the plot of the tear drop surface. The other programs for explicit and parametric equations have similar features. For a different view you need to replot the surface, but you have a great deal of control over this and the line density, so the output is high quality. To view slices, you just need to set the limits for plotting; for example to view the slice parallel to the yz plane through the point $x = 0.5$ use the range x-min $= 0.5$ and x-max $= 1$ or make x-max 0.51 to see it as a thin slice. The other programs in this suite, for explicit and parametric equations, have similar features.

DPGraph example

DPGraph is a shareware program (written by David Parker). It is very fast because it is written in assembly language, and has different kinds of control including the ability to rotate the surface in real time (although not with the accuracy of being able to enter rotation values as with Plot3D) and vary the values of variables using the scroll bar. It is also possible to set up the ability to move along the surface and slice using the scroll bar. However, this is only a tool to aid viewing rather than plot the output for modelling. I have found that setting a two dimensional view and plotting a thin slice is not sufficiently accurate for modelling.

DPGraph has the ability to plot more than one function at the same time, and to animate using a time variable and is very powerful for exploring surfaces more complex than would be suitable for Sliceforms. It complements Plot3D and data is entered as a set of text rather than through a dialog box, so the teardrop used in the Plot3D example above would look like this in its simplest form.

```
a := 0.4
a.minimum := -2
a.maximum := 2
b := 0.7
b.minimum := -2
b.maximum := 2
graph3d.box := false
graph3d.mesh := true
graph3d.view := standard
graph3d.perspective := true
graph3d.resolution := 80
graph3d.minimumx := -1
graph3d.maximumx := 1
graph3d.minimumy := -1
graph3d.maximumy := 1
graph3d.minimumz := -2
graph3d.maximumz := 2
graph3d(a*z^5 + b*z^4 - (y^2+x^2) = 0)
```

and adding shading through the menu, gives a view like this:

figure 10-8

GrafEq example

GrafEq is a shareware program (written by Pedagoguery Software in Canada) is a program for plotting all types of planar equations including implicit and parametric equations, with the added benefit that you can place constraints on the way the equation is plotted. It has a beautiful user interface which (as figure 10-9 shows) allows you to see the equations as you would write them by hand, rather than the semi-coded entry in Plot3D and DPGraph. GrafEq works by a process of successive refinement, so you see the curve grow before your eyes. It has many features which would allow even more surfaces to be designed; for example you can mix polar and cartesian coordinates, and use conditional expressions not just conventional equations. From the point of view of making Sliceforms, this offers simpler routes than the ones described later in the chapter in the section "Programming techniques for surfaces defined by conditions".

The following example shows the use of the equation:

$$x^4 + y^4 + z^4 - (x^2 + y^2 + z^2) = 0$$

which is a simplified version of the Goursat surface equation described below in the section "Creating a surface from an implicit equation".

When you use GrafEq to plot a function, you build up the algebraic relation and then plot in a view window. The algebraic relation window for this equation, with the constraint that x = 0 (that is plot the central y-slice) is shown in figure 10-9.

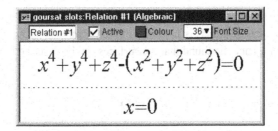

figure 10-9

You can plot a number of relations on the same graph, which is useful for adding the position of the slots. The view window in figure 10-10 also has such lines from y = 0 to y = 1 in steps of 0.25 plotted as well. This is simply a matter of adding a number of extra algebraic relationships and plotting them on the same graph. So, for example, the relationship y = 0.25 plots a vertical line.

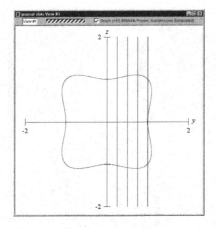

figure 10-10

The technique for creating the complete slices is described in the next section.

Creating a surface from an implicit equation

As described in chapter 7, plotting implicit equations has not been easy until fairly recently with software like that described above. The following example goes through the process of creating a model using this software. A more "creative" programming technique is described later in the chapter in the section "Programming techniques for surfaces defined by conditions".

This example shows a model of one of the surfaces studied by Goursat [see bibliography] each of which is invariant of the symmetry group of the regular polyhedra. For a cube, he found that the surfaces can be described by an equation of degree 4:

$$x^4 + y^4 + z^4 + a(x^2 + y^2 + z^2)^2 + b(x^2 + y^2 + z^2) + c = 0$$

Varying the values of the constants a, b and c in programs like Plot3D and DPGraph give interesting surfaces. You soon see that not all suitable for making Sliceform models, because they have holes which would cause the slices to break up. The examples in figure 10-11 have varying values of a, b and c, at the left, 0, -1 and 0.5 respectively; in the centre -0.125, -1 and 0.5; and at the right 0, -1 and 0. This dimpled cube is suitable for making a Sliceform model.

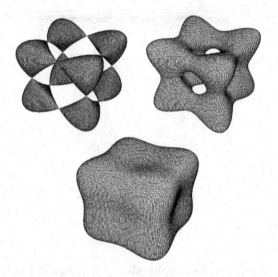

figure 10-11

After the initial stage of analysing the variables and ranges, the following technique describes how to make a model of this surface using the plotting software GrafEq. The cubic symmetry of the surface reduces the number of slices that have to be designed.

Stage 1 - plot slices

GrafEq allows you enter the equation directly as described above. The result is a pixel image and can be exported as such for further processing.

The obvious way to slice is parallel to a side of the cube. Keeping x constant gives a set of slices. A little experimentation with the value of x shows the effect of slicing as in figure 10-10:

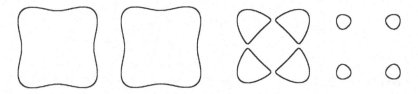

figure 10-12

The values of x are 0, .1, 1.1 and 1.15 going from left to right. Between 0 and 1 the indentations change shape. As seen in figure 10-12 as well as figure 10-13, once it goes above 1, then the slicing cuts off separated pieces and the limit of the solid is close to 1.15, so the range of slices with x having values 0, 0.25, 0.5, 0.75 and 1 would appear to give a suitable set.

Stage 2 - convert to vector images

Figure 10-12 has been produced as a vector image and not a pixel (bit) image. Design packages like CorelDRAW have an option to trace the image and produce a vector image. The resultant image can then be imported into the design module of CorelDRAW, but the following points need to adhered to:

- each slice must be plotted in the same size window (in this example 768 by 768 pixels);
- when each curve has been converted to a vector, they must *all* be imported onto the same page *before* any sizing takes place;
- when the equation is plotted, a reference is required for the dimensions; this can be achieved by displaying axes and zooming in to see where the curve cuts the axis.

The first two points ensure that the different slices remain the same relative size. The third point allows the positioning of the slots to be measured and is only needed for one slice. In this case, however, setting y and z to zero shows that the value of x equals ± 1. So the waist of the curve for the $z = 0$ slice is 2 units across. This means that although there is an obvious connected slice in figure 10-13 for a value of x of 1, it is the limiting case of a connected slice. Slicing the central slice at this position would cause it to separate as shown in figure 10-12, so only four slices are suitable.

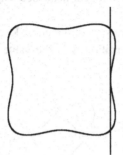

figure 10-13

When tracing the curve, you will get more accurate results in Corel Trace if you trace the centre line of the curve and also if you adjust the settings for highest accuracy. This means setting the Node Reduction to its lowest value and have at least 20 iterations.

Stage 2 - add the slots and assemble

The final stage is the same as for other models. As described earlier, design packages are not easy to work with when you need to measure. It is better to copy the vector slices into a CAD package for this purpose. This yields four separate types of slice, for x values 0, 0.25, 0.5 and 0.75. Two pieces are required for the x value 0 and four for the rest as shown in figure 10-14.

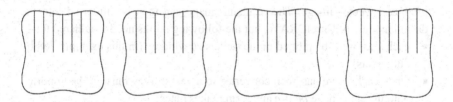

figure 10-14

These look very similar, but the Sliceform model shown in figure 10-15 shows a definite dimple when they are assembled.

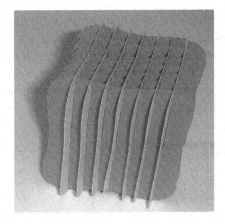

figure 10-15

Creating a surface from a parametric equation

Creating surfaces from parametric equations is very similar to the techniques
for implicit equations. The example of the cardioid based surface described in
chapter 7 and shown in figure 7-26 in that chapter, has been chosen show a
number of situations where you have to be careful in setting the parameters for
the programs you are working with. As with the implicit equation, the three
dimensional analysis needs a program like Plot3D or DPGraph. Further
analysis is required for GrafEq to ensure that the slices in each direction
remain at the same scale.

Stage 1 - plot slices

Entry of the equation in GrafEq is more detailed than with the implicit
equation described above. In effect it is a set of constraints for what you want
to plot. The relation window is shown in figure 10-16. Note that the three
parametric equations for x, y and z are given first, followed by the range for
the angles u and v and then the plotting is constrained to the plane x = 0.25.

cardiod based surface:Relation #1 [Algebraic]
Relation #1   Active   Colour   36 ▼ Font Size
$x=\sin(u)*(1+\cos(v))*\cos(v)$
$y=\sin(u)*(1+\cos(v))*\sin(v)$
$z=u$
$0<u<\pi$
$-\pi<v<\pi$
$x=0.25$

figure 10-16

When entering the range for the plot view window, there is also some analysis
required to find the range. It is also important that the size of the window used

for plotting the two sets of slices has the same absolute proportions, otherwise the resulting slices will not match even though their shape might be right. The best way to determine this is to plot slices in the xy plane (by editing the last constraint to give values of z). The maximum size is then when $z = \pi/2$. This view (figure 10-17) shows that the cardioid is such that slices are not the same in each direction and the origin is not central to the plot.

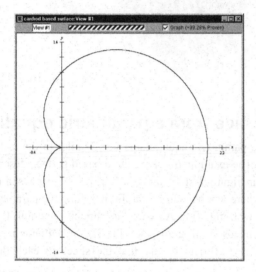

figure 10-17

Suitable ranges for plotting the slices are from -0.2 to 3.2 in both cases (since the height is the same) and -0.4 to 2.4 for the x direction slices and -1.4 to +1.4 for the y ones. This ensures that the absolute horizontal range plotted is 2.8 units in each case.

By plotting this central cardioid, it is also obvious that a slice along the y axis would not be ideal, so the number of slices in each direction in the model shown in figure 7-27 are seven in the y direction and nine in the x.

Stages 2 and 3 convert to vectors and assemble

The techniques for these stages are exactly the same as the ones for the implicit equations described above.

Programming techniques for surfaces defined by conditions

When you have no sophisticated tools like the ones described above, it is often necessary to be creative and invent your own. The following methods were ones I developed before having access to such software.

Figures 3-14 and 3-15 show a family of curves which use Maxwell's methods for their definition. They are defined as curves which meet a rule like the sum of the distance of a point of the curve from a set of foci is a constant. Such methods can obviously be extended to define surfaces in three dimensions. There are two approaches to this. One is to work out a complex implicit equation as for the three focus surface described in chapter 7 (see figures 7-13 and 7-14). The other way is to program the boundary of the slices of the surface and then use the techniques for tracing the edge of these curves as used for implicit equations plotted with GrafEqn as described above. The following programming method becomes easier the higher the number of focal points is considered. Thus the three focus surface in chapter 7 is manageable, but if you were to write the equation for an dodecahedron (with 20 points) it would tend to get either too complex or prone to error.

Our conception of curves and surfaces is defined both by the space we live in and the tools we use. Whereas we normally draw a free hand curve with a tool like pencil, it is not the curve we would describe mathematically, because the pencil draws a line of finite thickness. We can also think of it like a boundary or fence, particularly if it is a closed curve like a circle. So we can see a curve in figure 10-18, even though it has not been plotted as a line.

figure 10-18

The algorithm for the program is quite simple. Suppose we want to plot the slices for the surfaces where the sum or product of the distances to the points of a polyhedron are a constant. The steps for plotting a curve are:

1. Set up a screen which defines a slice (for example, choose a plane parallel to the xy plane having a fixed value of z).
2. For each pixel on the screen (with a specific x and y coordinate), calculate according to the rule for the sum or product to the polyhedral foci using Pythagoras' theorem.
3. If the value is above the chosen value for the constant then plot a coloured pixel.
4. Continue until the whole screen has been scanned and an image like figure 10-18 is obtained.

This technique could also be used for making the contours for the families of curves like figures 3-14 or 3-15. The curves must be built up one by one by plotting each one with a different value for the given length.

To convert the curve to a vector image, the process is similar to the one used for implicit equations above:

1. Use a paint program like Corel PHOTO-PAINT, Paintshop Pro or Adobe Photoshop, with an effects filter which is usually called something like "edge detect".
2. This gives an image which is like the plotted curve obtained from *GrafEq* as described earlier.
3. Follow the instructions in Stage 2 above for converting pixel plotted implicit equations into vector and adding the slots and assembling.

Example using this method

Using this technique is possible to create a new set of surfaces that do not appear to be present in the literature. The following example shows the surface defined by the rule that for each point on the surface the sum of the distances from the vertices of a cube are a given length. Figure 10-18 is a plot of one slice from such a surface produced as follows:

1. Define a set of slices for planes having fixed z values of 0, 0.3, 0.6, 0.9 and 1.2.
2. Set points on this plane with x and y taking values in the range ± 1.6 where the sum of the distances from eight coordinates of the cube of length 2 (that is coordinates $\pm 1, \pm 1, \pm 1$ are greater than or equal to 18.

All the slices look similar to the shape in figure 10-18, but vary in size. The Sliceform model is shown in figure 10-19.

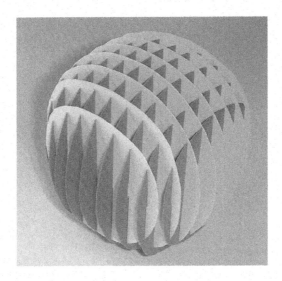

figure 10-19

Programming other devices

The programming methods described in this chapter describe how to produce images of slices which are then printed on paper. The slices are then cut out by hand. The next stage, which requires expensive equipment, is to program devices like laser cutters to cut out the slices for you. This overcomes the problems of having to transfer the outlines of the slices to thicker material such as wood or plastic. Such methods are beyond the scope of this book.

Where to find out more

Sliceforms is an evolving subject. The bibliography gives a background to the past and specific surfaces or the mathematics behind them, together with some software for downloading. While the book will not be out of date as a store of knowledge and concepts, new ideas come up for different models or their use (for example in teaching). The internet is the source of new information, so regular searching is one way to find out what is going on.

The second way is to look at the Sliceforms user section on *www.mathsite.co.uk* which is the QED website. There, as this book goes to press, you will be able to find:

1. Links to Sliceforms on the internet.
2. Templates to download.
3. A gallery of models, real and virtual, including full colour version of many of the images in this book.
4. A forum for exchange of ideas.
5. New books and software, many of which you can buy from the site should you so wish.

Please join in and show what you are doing with Sliceforms, how you are using them and any new models you have made.

Bibliography

E T Bell, "Men of Mathematics"
A series of historical portraits of mathematicians.

Robert J T Bell, "Coordinate Geometry of three dimensions", Macmillan, 1931
Good general book on three dimensional coordinate geometry. Includes slicing
of the torus p 267.

Masahiro Chatani, "Origamic Architecure", Ondori Publications 1984
One of a series of books by Chatani, who has combined his architecural skills
with origami to produce pop-up cards which include Sliceform variations.

H S M Coxeter, "Introduction to Geometry", Wiley, 1969
Contains a useful introduction to inversion and discusses the slicing of a torus
(p132).

H M Cundy and A P Rollett, "Mathematical models", Tarquin Publications.
The classic work on making mathematical models in all dimensions which,
because it is getting dated, includes a great deal of historical material.

Walther von Dyck "Katalog mathematischer und mathematisch-physilaischer
Modelle, Apparate und instrumente." 1892-3, reprinted 1994
Contains many woodcuts of the models for sale.

Gerd Fischer, Mathematical Models Vieweg & Sohn Braunschweig 1986
The first volume could be considered a coffee table with beautiful photographs
of the models. The second volume is a commentary containing the
mathematical background.

Martin Gardner, "Mathematical Carnival", Pelican 1978
The chapter entitled "Piet Hien's Superellipse" describes the properties of
superellipses and the superegg. The addendum to the chapter has interesting
information on balancing the superegg.

P G Gasson, "Geometry of spatial forms", Ellis Horwood, 1983

Written to provide background geometry for computer aided design of solids and surfaces, this provides useful information if you need to work with equations for surfaces.

E Goursat, "Etude des surfaces qui admettent tous les plans de symétrie d'un polyedre régulier." Ann. Sci Ec. Norm, supp (3) **4** p 159-200, 1887 (see also Fischer chapter 2 and photo 48)

D. Hilbert and S. Cohn-Vossen, "Geometry and the Imagination", Chelsea 1952
Originally published in German in 1932 as "Anschaulische Geometrie" which was based on a course of lectures given by David Hilbert in 1920-21 at the University of Göttingen and is one of the classics of mathematics. Hilbert was one of the giants of mathematics in the early part of the century.
It is a very readable book with a minimum of equations, but a host of illustrations, many of which are photographs of the models at Göttingen. A large part of the book deals with the geometry and topology of surfaces.

Tord Hall, "Carl Friedrich Gauss, a biography" MIT Press 1970
Contains a description on Gauss's theory of curvature of surfaces.

S Hildebrandt and A Tromba, "Mathematics and optimal form", Scientific American Books, 1985
A very readable book on surfaces which is strong on history. Very well illustrated and particularly strong on soap bubble and other minimal surfaces in nature.

Alan Holden, "Shapes, Space and Symmetry", Columbia University Press 1971, reprinted Dover 1999.

T Kiang, "An old Chinese way of finding the volume of a sphere", Mathematical Gazette, Volume LVI (396), May 1972, pp 88-91.
This is an ingenious method of relating volumes using slices and is well worth investigating if you are interested in the mathematics of slicing solids.

E A Lord and C B Wilson, "The mathematical description of shape and form", Ellis Horwood 1984
In places the mathematics is very advanced, but in others reliance is made on diagrams, so well worth looking at for inspiration.

E H Lockwood, *"A book of Curves"* CUP 1961
One of the standard books on drawing curves.

Edward Lucie-Smith, "Wendy Taylor", Art Books International, 1992.
Wendy Taylor's sculpture is strongly influenced by mathematics. One of her pieces is a Sliceform. Others are based on the Mobius strip and knots.

W H McCrea, "Analytical Geometry of three dimensions", Oliver and Boyd 1953.
Briefly shows how to make an ellipsoid with circular sections.

Elisabeth Mühlhausen, "Riemann Surface - Crotcheted in Four Colours", Mathematical Intelligencer 15 (3), 1993, pp 49-53
Part of the "Mathematical Tourist" series, a tour of the Mathematical Institute museum of the University of Göttingen with many photographs of models. No Sliceforms are visible.

C Stanley Ogilvy, "Excursions in Geometry" originally published 1969 by Oxford University Press, republished Dover 1990.
An interesting book on recreational geometry, with a number of chapters dealing with inversion in the plane.

Ivars Peterson, "The Mathematical Tourist" Freeman, 1988

Melvin Prueitt, "Computer Graphics: 118 Computer Designs", Dover
Shows the use of mathematical surfaces as art objects in early computer art by one artist.

Jasia Reichardt, "The Computer in Art", Studio Vista, 1971
A survey of computers in art when they were just beginning to make an impact. The work of Charles Csuri was strongly influenced by mathematical models of surfaces and is shown in a number of works.

W von Rönik, "Doughnut slicing", College Mathematics Journal, vol 28, Nov 1997
Investigation of the slicing of a torus to give the Villarceau circles.

Martin   Schilling   "Mathematiche   Abhandlung   aus   dem   Verlage
Mathematischer Modelle" Halle 1904
At one time the largest retailer of mathematical models. Contains many
woodcuts of the models for sale.

Science and Art Department of the Committee of Council on Education,
"Catalogue of the Special Loan Collection of Scientific Apparatus at the South
Kensington Museum 1876".
The catalogue has a description of the surfaces and geometrical models
exhibited. The entry included in the section on the background to the models is
a typical example of the detail included.

D Seggern, "CRC Handbook of Mathematical Curves and Surfaces"
A survey of curves and surfaces which is lavishly illustrated with both pictures
and equations for surfaces. The second edition has better illustrations and
more surfaces.

John Sharp, "Sliceforms, Mathematical models from paper sections", Tarquin
Publications, Diss 1995

John Sharp, "Sliceform craters, an exploration in equations", Mathematics in
School March 2004

Eugene V Shikin, "Handbook and Atlas of Curves", CRC Press, 1995

D Y Sommerville, "Analytical Geometry of Three Dimensions", Cambridge,
1934
The mathematics of the deformation of the Sliceforms of the quadric surfaces
is described on pages 204 to 207. He also describes why they can be built from
circular sections.

Tate Gallery, "Naum Gabo: Sixty Years of Constructivism", 1987
Contains a detailed chapter on the influence of mathematics on art before the
Second World War. This is particularly relevant to the influence of Sliceforms
on Gabo's sculpture.

H W Turnbull, "Collapsible circular sections of quadric surfaces", Edinburgh
Maths Notes, No 32, 1944, pp xvi-xix.
Describes the mathematics of Sliceforms as a method of teaching the use of
oblique axes.

Laurence Weider, "Solid Geometry", Camera Arts July 1983 pp 30-37, 77
Describes (with some photographs) models in the Institute Poincaré. The artist
Man Ray was commissioned to photograph them as *objet d'art* when they were
rediscovered there in the 1930s by Max Ernst.

David Wells {1}, "You are a Mathematician" Penguin 1995.
More two dimensional, but has some interesting ideas on maps which may
give you some ideas.

David Wells {2} (illustrated by John Sharp), "The Penguin Dictionary of
curious and interesting geometry", Penguin 1991, 0-14-011813-6
The title says it all. There are many surfaces described.

R C Yates, "A handbook on curves and their properties", Ann Arbor 1947,
Reprinted by the NCTM.
Although this covers curves, the methods suggest ways in which new surfaces
can be designed and offer ways to construct curves for surfaces of revolution

Antoine J F Yvon Villarceau, Nouvelles Annales de Mathematiques (1), vol 7,
1848, pp 345-47
This paper describes the discovery of the Villarceau circles of the torus. See
chapter 9.

Specialised Software

DPGraph (Dynamic Photorealistic Graphing)
A shareware package by David Parker, which graphs from two dimensions
upwards and is very fast. It can cope with explicit, implicit and parametric
equations and allow you to vary parameters in real time.
Available at www.dpgraph.com

* *GrafEq* from Pedagoguery Software is an exceptional tool for exploring
  mathematics producing two dimensional plots from a wide range of input
  formats which can be entered in an intuitive way.

Available at http://www.peda.com/grafeq/

Plot3D is a set of programs for plotting surfaces by Clark Dailey, with data
entered through a dialog box and allowing very fine control over the plotting.

Available at `www.simtel.net` and also on CD from Simtel. Download the files from the Win98 section / Graphics /Misc. Graphic Programs & Utilities There are three programs in the suite.

Plot3DF3.ZIP - Plot F(x,y,z)=0 as 3D surface
Plot3DP1.ZIP - Plots a 3D surface parametrically.
Plot3DZ3.ZIP - Plot z=f(x,y) as 3D surface

Acknowledgements

Thanks to Rhys Williams for the idea and photographs of the chair in chapter 9. Thanks also to Richard Ahrens for the photographs of his hyperbolic paraboloid sculpture in chapter 6. Finally thanks to Richard and Sue Ahrens for their help and advice in getting the book to its final form. The responsibility for any remaining errors entirely mine.

Using the templates

The following pages consist of series of templates for making your own Sliceform models. They have been chosen to show a range of types of surface and modelling techniques. Follow the instructions below for each model.

Important points

1. You should be familiar with chapter 1 before attempting to make these models. The methods for cutting and assembly are described there with **the instructions for cutting the slots being particularly important**.
2. The templates are meant to be photocopied onto card. Some models require enlargement. Each page needs to be photocopied twice to get the slices for each direction. This means that some models will then have duplicate slices. See the instructions for each model for details.
3. Some sheets have two models on them.

The crater

This model is described in chapter 7.

You should make two copies without enlarging. When you have made the copies, you will have an extra version of the central slices (slices X1 and Y1), since slices for both directions are on the sheet. The X slices have slots coming from the top and the Y slices from the bottom. Cut the card so that you have two sets of slices X2 to X5 and Y2 to Y5. Cut out the central slices (X1 and Y1) and slot them together. Then cut a pair of X2 and a pair of Y2 and add them to the model either side of the central slices. Continue with the next pairs until you have finished.

Ellipsoid as a spheroid

This model is described in chapter 2.

You should make two copies enlarging by around 130% to make the largest, central slice about 8 cm. Because the surface is symmetrical, the slices are the same in both directions. Cut out the central slices (slice 1) and slot them together. Then cut a pair of slice 2 from each sheet and add them to the model either side of the central slices. Continue with the next pairs until you have finished.

Superegg

This model is described in chapter 5.

You should make two copies enlarging by around 125% to make the largest, central slice about 6.5 cm. Because the surface is symmetrical, the slices are the same in both directions. Cut out the central slices (slice 1) and slot them together. Then cut a pair of slice 2 from each sheet and add them to the model either side of the central slices. Continue with the next pairs until you have finished.

Hyperbolic paraboloid

This model is described in chapter 6.

The template does not need enlarging. A complete set of slices is given, so if you want to make a model with different coloured slices in each direction, then make two copies on different card and make two models. The same is true for the tetrahedron on the same page.

When building the model, slot the central two slots of the two rectangles together and then add two of the other pieces (one from each direction) together as in figure 18 in chapter 5. Because of the position of the slots, this requires weaving the slices so that the model does not fall apart. When adding the outside slices, ensure that the symmetry at the top corner is correct to give the butterfly wings effect as in the figure in chapter 5.

Small tetrahedron

This model is described in chapter 8.

The template does not need enlarging. A complete set of slices is given, so if you want to make a model with different coloured slices in each direction, then make two copies on different card and make two models. The same is true for the hyperbolic paraboloid on the same page. If you are making two models, then you can try making them with mirror symmetry as shown in figure 17 in chapter 9.

When building the model, slot the central two slots of the two squares together and then add the other pieces so that the two outside pieces in the same direction are rotated 90 degrees. Look at the diagrams and photos in chapter 9 if you are having trouble.

Monkey saddle

This model is described in chapter 7. This model shows how even a small number of slices can be used to model a surface.

The template does not need enlarging. A complete set of slices is given, so if you want to make a model with different coloured slices in each direction, then make two copies on different card and make two models.

The number of the slices is from front to back for the X slices, and left to right for the Y slices (although the outside ones are identical). Cut out the central slices (X2 and Y2) and slot them together. Then cut out the others and add them to the model either side of the central slices.

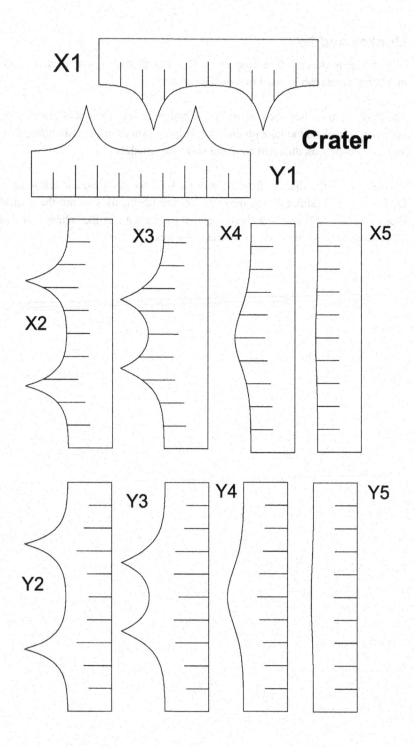

X1

Crater

Y1

X3   X4   X5

X2

Y3   Y4   Y5

Y2

Ellipsoid as spheroid

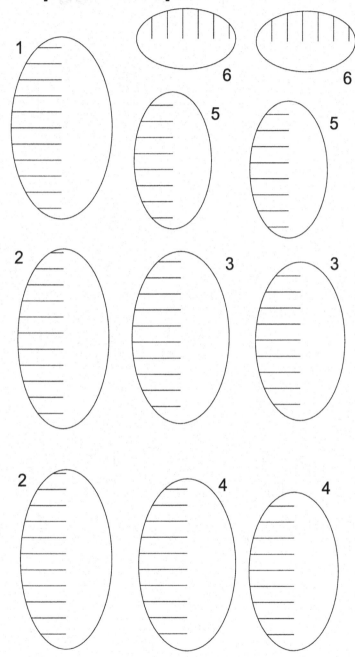

Superegg

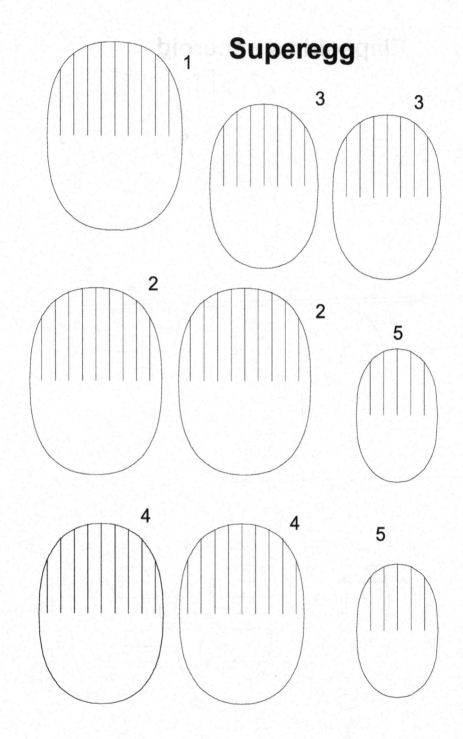

Hyperbolic paraboloid

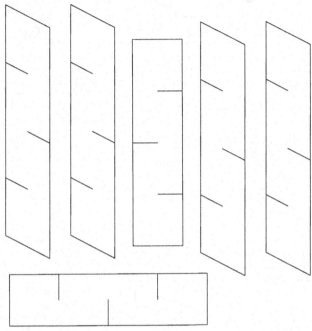

Small tetrahedron

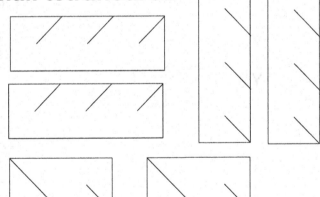

Monkey Saddle surface

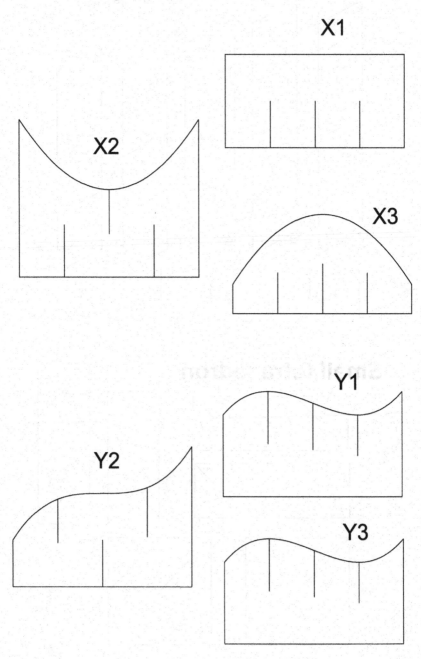

Index